ARROYO CENTER

T0288594

# Improving Strategic Competence

## Lessons from 13 Years of War

Linda Robinson, Paul D. Miller, John Gordon IV, Jeffrey Decker,
Michael Schwille, Raphael S. Cohen

Prepared for the United States Army

For more information on this publication, visit www.rand.org/t/rr816

Library of Congress Cataloging-in-Publication Data is available for this publication.

ISBN 978-0-8330-8775-1

Published by the RAND Corporation, Santa Monica, Calif.
© Copyright 2014 RAND Corporation
RAND® is a registered trademark.

### Support RAND
Make a tax-deductible charitable contribution at
www.rand.org/giving/contribute

www.rand.org

# Preface

The purpose of this report is to contribute to the ongoing efforts to distill lessons from the U.S. experience in 13 years of war (2001–2014), apply the lessons to the future operating environment, and identify critical requirements for land forces to operate successfully in conjunction with joint, interagency, and multinational partners to address hybrid and irregular threats. This study seeks to address a particular gap in the current debate on the future of national security strategy and the role of landpower caused by an inadequate examination of the national level of strategy made by the interagency level of government. The gap exists because there has been no systematic effort to collect and analyze insights from those who have been actively engaged in making policy and strategy from 2001 to 2014. A RAND Arroyo Center workshop provided a mechanism for eliciting insights from policymakers and academic experts involved in the formation of national-level strategy and its implementation over the past 13 years. This study analyzes and develops those insights in the context of the overall debate on future national security strategy. The purpose of this document is to assist military and civilian leaders in assessing capabilities needed in the U.S. government, and in land and special operations forces in particular, in future irregular and hybrid conflicts.

This research was sponsored by U.S. Army Special Operations Command and conducted within the RAND Arroyo Center's Strategy, Doctrine, and Resources Program. RAND Arroyo Center, part of the RAND Corporation, is a federally funded research and development center sponsored by the United States Army.

The Project Unique Identification Code (PUIC) for the project that produced this document is HQD146692.

# Contents

# Tables

# Summary

The United States and many of its closest allies have been engaged in a long period of continuous military operations with mixed success in confronting a range of complex and dynamic threats. The U.S. military recognizes that a great deal of intellectual work remains to be done to learn from these experiences. This study seeks to contribute to the ongoing debate about the lessons from the past 13 years of war and the requirements for addressing future conflicts. It addresses a particular gap in the current debate on the future of national security strategy and the role of landpower caused by an inadequate examination of this recent experience in the national-level of strategy made by the interagency level of government. The gap exists because wartime pressures have not afforded time for reflection by many of those who have been actively engaged in making policy and strategy for the past 13 years and because there has been no systematic effort to collect and analyze such insights. The lessons collected and analyzed here represent an initial contribution to the ongoing collective endeavor to learn from these recent conflicts.

The U.S. military leadership has conducted its own examinations and recognized the need for continued study of the recent past. At the direction of the Chairman of the Joint Chiefs of Staff, in June 2012 the Joint and Coalition Operational Analysis (JCOA) division of the Joint Staff J-7 produced an initial study examining recent wartime experience. That study, *Decade of War, Vol. 1: Enduring Lessons from the Past Decade of Operations*, identified 11 broad themes, based on 46 previous JCOA studies analyzing U.S. operations in Iraq, Afghanistan, Libya, and elsewhere.[1] In addition, in a white paper published in May 2013, "Strategic Landpower: Winning the Clash of Wills," the U.S. Army, the U.S. Marine Corps, and the U.S. Special Operations Command announced the formation of a task force committed to "a thorough study of how all the elements of landpower can be best employed to support national strategic objectives." The paper emphasized the need for further examination of the recent wartime experience, based on the proposition that "when we have formally studied the relevant lessons of our past, and applied that knowledge against the risks posed by the future operating environment, we have come away better postured to advance or achieve our nation's strategic objectives." The endeavor specifically sought to "help inform the Defense establishment's thinking on better integrating human factors into the planning and execution of military operations to achieve enduring outcomes."

This RAND Arroyo Center report builds on the joint staff's examination and the task force's work with an expanded scope that includes policy-level issues and interagency perspectives. The lessons formulated on the basis of a RAND Arroyo Center workshop and subsequent analysis are then applied to the future operating environment, which includes irregular and hybrid threats, in order to identify critical requirements for land and special operations forces (SOF) to operate successfully in conjunction with other joint, interagency, intergovernmental, and multinational partners.

The approach employed in this study consisted primarily of document-based research and semistructured interviews with experts and officials involved in the making and implementation of policy and strategy in the past 13 years of war. In addition, the RAND Arroyo Center study team convened a workshop of policymakers and academic experts in national security, civil-military relations, and strategy to discuss the policy, strategy, and implementation lessons of the past 13 years. The workshop was followed by a two-stage Delphi poll to determine areas of agreement among the participants.

---

[1]  JCOA, *Decade of War, Vol. 1: Enduring Lessons from the Past Decade of Operations,* June 15, 2012.

To place the past 13 years of war in historical context, Chapter Two examines the U.S. experience in warfare from World War II to the present and identifies broad trends and inflection points in how U.S. forces have fought. Two themes emerge from this survey. First, land warfare has evolved away from conventional combat against state actors and their standing forces to an increasing incidence of irregular warfare fought by joint forces against nonstate actors. This has led to an increasing reliance on SOF, which have grown and participated in a wider range of military operations than at any time in their history. Second, while the Army often learns tactical and operational lessons from the wars it fights, it often struggles to incorporate these wars' broader strategic lessons that require it to think outside of the box and beyond its immediate past experiences. Thus, the Army and the joint force as a whole have adopted new technologies that have improved the mobility, survivability, and situational awareness of forces; the ability to operate at night; and the lethality and precision of weaponry. Yet the joint force and the U.S. government as a whole have displayed an ongoing ambivalence about and lack of proficiency in the noncombat and unconventional aspects of war and conflict against nonstate actors, despite their increasing frequency. Much of the past 13 years were devoted to recovering, refining, and partially institutionalizing those capabilities. The challenge now is to preserve and refine needed capabilities and develop innovative new ways of operating based on recent experience.

Chapter Three formulates seven lessons from the past 13 years of war based on research, interviews, the workshop, and the Delphi exercise. Many of these topics predate the past 13 years and have been the subject of extensive debate, scholarship, and attempted policy remedies. The seven lessons identified in this study were determined to be relevant to both the recent past and the future. They are as follows:

Lesson 1. A deficit in the understanding of strategy
Lesson 2. Deficits in the process for formulating strategy
Lesson 3. A failure to incorporate the essential political element of war into strategy

Lesson 4. The inability of technology to substitute for the sociocultural and historical knowledge needed to inform understanding of the conflict, formulation of strategy, and timely assessment

Lesson 5. A failure to plan, prepare, and conduct stability operations and the transition to civilian control, as well as belated development of counterinsurgency capabilities

Lesson 6. Insufficient emphasis on shaping, influence, and noncombat approaches to addressing conflict

Lesson 7. Inadequate civilian capacity and inadequate mechanisms for coordinated implementation among joint, interagency, and multinational partners.

The lessons are described in greater detail below, with additional argument and supporting evidence provided in Chapter Three.

*Lesson 1: The making of national security strategy has suffered from a lack of understanding and application of strategic art.* The U.S. government has experienced a persistent deficit in understanding and applying strategic art. The blurry line between policy and strategy requires both civilians and the military to engage in a dynamic, iterative dialogue to make successful strategy, but that often failed to occur. The decision to go to war in Iraq, the decisions to send a surge of troops to Iraq and then Afghanistan to bolster faltering war efforts, and the approach taken toward countering terrorism in the past two administrations all illustrate strategy deficits. In the first case, the civilian policymakers did not seek and factor in the needed information to examine their assumptions and prepare for likely consequences. In the second case, the civilian policymakers found the military's recommendations inadequate and relied on outside advice in making the decision to surge in Iraq. In the case of the Afghan surge, that decision was reached after multiple reviews stretching over two years, but it did not resolve the divergence in approach favored by the military (full-spectrum counterinsurgency) and senior civilians in the White House who advocated a narrower counterterrorism agenda aimed at al Qaeda. The fourth major decision, ratified and pursued by both administrations, was to adopt a global counterterrorism strategy that relied primarily on strikes against terrorists who were actively plotting to strike the United States. During the past 13 years, the strategies typically failed to envision a war-ending approach and did not achieve declared objectives in a definitive or lasting manner. The ends, ways, and means did not align, whether because the policy objectives were too ambitious, the ways of achieving them ineffective, or the means applied inadequate.

*Lesson 2: An integrated civilian-military process is a necessary, but not sufficient, condition of effective national security strategy.* The current process does not routinely produce effective strategy, in part because the U.S. military is taught to expect a linear approach in which the policymakers provide the objectives and the military develops the options for achieving them. The military typically provides a range of operations but prefers one that meets the objectives fully with the least risk. That model falls short in two respects. Civilian policymakers require an active dialogue with the military and other sources of information to inform the diagnosis of the situation, as well as to develop realistic policy objectives. That iterative process must continue through the development of options, since the president weighs a wide variety of factors in considering the optimal course of action. Formulating strategy is further inhibited because there is no established integrated civilian-military process that would rigorously identify assumptions, risks, possible outcomes, and second-order effects through soliciting diverse inputs, red-teaming, and table-top exercises. The lack of such a process inhibited timely adaptation of strategy in response to the evolution of understanding and events.

*Lesson 3: Because military operations take place in the political environment of the state in which the intervention takes place, military campaigns must be based on a political strategy.* Lesson Three examines the failure to think in terms of the political aspects of a conflict and the desired outcomes that are fundamentally political in nature. This deficit results in part from a tendency to focus on tactical issues, troop levels, and timelines, rather than the strategic factors that will determine a successful outcome. The U.S. military has also been reluctant to grapple with the political aspect of war, in the belief that it is either not part of war or entirely up to the civilians to address. Yet an intervention is unlikely to produce lasting results without a strategy that addresses the political factors driving the conflict and provides for enduring postwar stability. Implementing that strategy is likely to involve a combination of military and political means by the United States and local partners acting in concert—such as elections, negotiations, and power-sharing. This fundamental issue has been obscured by the focus on governing capacity, which is a separate, long-term, institutional issue that is often secondary to resolving conflict.

*Lesson 4: Because of the inherently human and uncertain nature of war, technology cannot substitute for sociocultural, political, and historical knowledge and understanding.* Lesson Four finds a deficit of both sociocultural and historical knowledge that is critically needed for understanding a conflict, formation of strategy, and the assessment of its implementation. In part, this is due to a continuing overreliance on technology and a belief that wars can be fought and won by reliance on it alone. Without such knowledge and understanding, necessarily developed over time, the required adaptations in the strategy cannot be made. While the need for assessment has been acknowledged, the approach to assessments may rely too heavily on systems analysis and on creating a process that charts the execution of a strategy (important but not sufficient) rather than its effect on the conflict.

*Lesson 5: Interventions should not be conducted without a plan to conduct stability operations, capacity-building, transition, and, if necessary, counterinsurgency.* Lesson Five finds that there was repeated failure to plan, prepare, and generate adequate capability and capacity for stabilization and reconstruction operations, capacity-building, and transitions to civilian authority (known respectively as Phase IV and Phase V operations in military doctrine) and conduct those operations in a sufficient and timely manner following interventions (Phase III major combat operations). Doing so in conjunction with a political strategy might have lessened or obviated the need for large-scale counterinsurgency operations. Counterinsurgency capabilities were developed but somewhat belatedly. Transitions to civilian authority were also hobbled by a failure to plan and implement the needed measures with the needed authorities in a timely fashion.

*Lesson 6: Shaping, influence, and unconventional operations may be cost-effective ways of addressing conflict that obviate the need for larger, costlier interventions.* Lesson Six finds that there is a chronic lack of emphasis on shaping, influence, and unconventional approaches that might in some cases avoid the need for Phase III major combat operations. The lack of emphasis can be traced to (1) a reluctance to engage in a proactive manner while a conflict is still relatively small or unthreatening, (2) an insufficient understanding of the full range of possible activities, and (3) an underdeveloped model for planning and conducting these operations as a campaign that achieves results without major combat. SOF have demonstrated the ability to achieve effects through a sustained campaign approach, in conjunction with other joint, interagency, and multinational partners, as an alternative to major combat operations. Yet the paradigm is not fully established, as "Phase 0" shaping, influence, capacity-building, and unconventional activities are often seen as a prelude to and preparation for major combat operations rather than a potential alternative to them.

*Lesson 7: The joint force requires nonmilitary and multinational partners, as well as structures for coordinated implementation among agencies, allies, and international organizations.* Lesson Seven finds that despite the issuance of directives to establish the necessary capability, there is a chronic lack of civilian capacity, as well as obstacles to civilians operating in hostile environments. Despite some successes in coordinated implementation among joint, interagency, intergovernmental, and multinational partners, the mechanisms for achieving the desired synergy are still inadequate, including in circumstances where a civilian-led effort is most appropriate.

Chapter Four examines characterizations of the future operating environment to determine whether and how the lessons of the recent past may retain their relevance. It concludes with recommendations for development of a theory of success and further adaptation and preservation of capabilities.

The National Intelligence Council's (NIC's) *Global Trends 2030: Alternative Futures* report finds that irregular and hybrid warfare will remain prominent features of the future threat environment. It states that "most intrastate conflict will be characterized by irregular warfare—terrorism, subversion, sabotage, insurgency, and criminal activities" and that intrastate conflict will also be increasingly irregular, noting that "[d]istinctions between regular and irregular forms of warfare may fade as some state-based militaries adopt irregular tactics."[2] The NIC report and others, such as the National Review Panel report on the Quadrennial Defense Review, also foresee an increased incidence of hybrid warfare caused by the diffusion of lethal weaponry and other factors that empower adversaries. If these projections are accurate, the need to heed the lessons of the past 13 years remains urgent, in order to preserve and refine the relevant capabilities and insight developed in these years, as well as to remedy those remaining gaps.

U.S. national security strategy has begun to grapple with this passage from the recent past to a future that will be resource-constrained yet still rife with many challenges to U.S. interests around the world. Defense strategic guidance has posited a rebalancing and rationalization of the joint force based on a force-sizing construct of fighting and winning one major war while denying victory in a second conflict; this represents a significant departure from the previous construct, which held that the United States needed to be able to fight and win two major wars, even if not concurrently (i.e., "win-hold-win"). The current defense strategic guidance states that the U.S. military will not size the force to conduct large-scale counterinsurgency and stability operations but that it will maintain the expertise and the ability to regenerate the needed capacity.[3] The joint force is currently in the process of determining what those two requirements entail in terms of needed capabilities.

The rise of irregular threats and constraints on resources pose an acute dilemma for U.S. strategy, increasing the imperative to remedy the deficiencies of the past 13 years. More than ever, the United States requires new approaches that can achieve satisfactory outcomes to multiple, simultaneous conflicts at acceptable cost. It must become more

---

[2]   National Intelligence Council, *Global Trends 2030: Alternative Worlds*, NIC 2012-001, December 2012, pp. 59–60.

[3]   Quadrennial Defense Review, 2014. Page VII states that "Although our forces will no longer be sized to conduct large-scale prolonged stability operations, we will preserve the expertise gained during the past ten years of counterinsurgency and stability operations in Iraq and Afghanistan. We will also protect the ability to regenerate capabilities that might be needed to meet future demands." It also states that "[t]he Department of Defense will rebalance our counterterrorism efforts toward greater emphasis on building partnership capacity."

agile in adapting its strategy as circumstances warrant, and it must improve its ability to work effectively with all manner of partners. The growing role of SOF represents a potential advantage of strategic import, but operational concepts and constructs must be further refined to supply a seamless array of options for the application of joint, interagency, intergovernmental, and multinational power.

Chapter Four also examines potential remedies that would apply the lessons identified in Chapter Three to future conflict. To address the strategy deficit and to provide a basis for determining the capabilities needed to address the irregular and hybrid threats of the future, Chapter Four advances the argument that a "theory of success" would provide a compass for strategy, address the full dimensions of war, and provide the basis for developing a wider array of effective approaches to resolve or contain threats. This chapter refers to theoretical and historical antecedents to posit a more robust conception of political warfare and political strategy as integral to U.S. national security policy and strategy. This view connects war and statecraft on the same spectrum and depicts the exercise of power as a marriage of force and diplomacy that wields the various elements of national power in a more seamless manner. The study concludes with seven recommendations for further adaptation and the retention or refinement of numerous capabilities:

- First, it recommends enhancing strategic competence by educating civilian policymakers and revising the version of policy and strategy taught to the U.S. military. It recommends adoption of an integrated civilian-military process that provides the needed expertise and information to diagnose the situation and formulate reasonable objectives, as well as subsequent strategy. Two options based on the Eisenhower National Security Council process are suggested.
- Second, the U.S. military may profitably explore deeper organizational transformations to increase its adaptability. Specifically, the military should examine ways to build effective, tailored organizations that are smaller than brigades and equipped with all the needed enablers to respond to a range of contingencies. This may entail significant institutional reform.

- Third, SOF and conventional forces should expand their ability to operate together seamlessly in an environment of irregular and hybrid threats. The recent robust use of SOF suggests the possibility of a new model, or models, for achieving operational or even strategic effect through a campaign approach. This represents a potentially potent new form of landpower that, if applied with strategic patience, can address threats without resort to large-scale military interventions. SOF-led campaigns can provide low-visibility, high-return security solutions in numerous circumstances. SOF have begun to develop the operational level art, planning, and command capabilities to realize this potential, but several additional steps are needed. In particular, new operational-level command structures can facilitate both SOF-centric and SOF-conventional operations. Habitual SOF-conventional teaming will maintain and deepen the interdependence and familiarity gained in the past decade. Reopening the advisory school at Fort Bragg can be a powerful mechanism for developing common procedures and understanding for operating in small, distributed, blended formations, as well as a ready cadre of trained advisers able to meet the expected demand of a national security strategy that places increased emphasis on partnered operations and building partner capacity.

- Fourth, innovative and multifunctional personnel can make a smaller force more effective, but the incentives must be systemic to reward personnel for creativity, risk-taking, and acquisition of multiple specialties. The principle of mission command can be deepened to permit further decentralization and delegation of initiative.

- Fifth, joint and service capabilities that create and maintain regional familiarity or expertise, advisory capability, and other special skills for irregular warfare and stability operations should be preserved and refined at the level needed to execute current military plans. These personnel can serve as a training cadre for rapid expansion in the event of a large-scale stability operation or counterinsurgency. The same skills are fungible in Phase 0 shaping and influence operations, so the cadre will likely be in high demand.

- Sixth, civilian expertise is essential in a broadened conception of war that places due emphasis on the political dimension. The most valuable contribution that civilians can make is often their expertise and insights, rather than hands-on execution at a tactical level. Because civilian capacity is likely to remain limited, the emphasis should be on ensuring that the relevant civilian experts are collocated at the key commands and sufficiently robust country teams and, when necessary, at the tactical level in formations that provide force protection and enable them to perform their duties.

- Seventh, multinational partners have proven their value in numerous ways over the past decade, but the U.S. government can improve its preparation of U.S. personnel to serve in coalitions and to effectively employ non-U.S. expertise by identifying in a systematic manner both its own gaps and the potential external resources to meet them.

This study identifies critical lessons from the past 13 years of war and recommends that a deliberate effort be undertaken to remedy the deficits in the "American way of war"; to preserve and improve the ability to tackle the strategic, political, and human dimensions of war; and to explore innovative new combinations of SOF and conventional forces to anticipate and meet the security challenges of the future more successfully.

# Acknowledgments

We are indebted to Michèle Flournoy, James Dobbins, Eliot Cohen, Frank Hoffman, LTG H.R. McMaster, James Schear, Janine Davidson, Charles Ries, Peter Feaver, Ciara Knudsen, Andrew Liepman, Vikram Singh, Paul Stares, Joseph Collins, Lincoln Bloomfield Jr., Nadia Schadlow, Beth Cole, Audrey Kurth Cronin, John Herbst, Stephanie Miley, John McLaughlin, Kevin O'Keefe, Richard D. Stephenson, CAPT John Burnham, Lt. Col. Robert Smullen, and Robert Giesler for their insights and relevant research. In addition, we are grateful to Arroyo Center director Terrence Kelly and our RAND colleagues David Johnson and Karl Mueller for their insightful comments and to Charles Ries, Adam Grissom, and David Maxwell for careful formal reviews of the draft report and numerous recommendations that greatly improved the final product. Peter Wilson provided invaluable research and insight on weapons technology and the future of warfare. We thank Sean Mann, Amanda Hagerman-Thompson, Maria Falvo, Terri Perkins, and Todd Duft for their valuable assistance and Nora Spiering for her professional editing of the document. COL Tim Huening, Larry Deel, and Jim Lane of U.S. Army Special Operations Command (USASOC), supported our work throughout this endeavor. Finally, the sponsor, USASOC commanding general LTG Charles T. Cleveland, spent many hours discussing emerging concepts and past experiences that helped inform the project team's work. Any errors or omissions are entirely the responsibility of the authors.

# Abbreviations

| | |
|---|---|
| AFRICOM | U.S. Africa Command |
| AQ | al Qaeda |
| AQI | al Qaeda in Iraq |
| ASG | Abu Sayyaf Group |
| ATGM | anti-tank guided missile |
| BCT | Brigade Combat Team |
| CIA | Central Intelligence Agency |
| COIN | counterinsurgency |
| CORDS | Civil Operations and Revolutionary Development Support |
| CT | counterterrorism |
| DoD | U.S. Department of Defense |
| FID | foreign internal defense |
| ISAF | International Security Assistance Force |
| ISR | intelligence, surveillance, and reconnaissance |
| JIIM | joint, interagency, intergovernmental, and multinational |
| MANPADS | man-portable air defense systems |

| | |
|---|---|
| MNF-I | Multinational Forces—Iraq |
| NSC | National Security Council |
| NCW | network-centric warfare |
| PRT | Provincial Reconstruction Team |
| QDR | Quadrennial Defense Review |
| RAF | regionally aligned force |
| RMA | revolution in military affairs |
| SOCOM | U.S. Special Operations Command |
| SOF | special operations forces |
| SOFA | Status of Forces Agreement |
| SRAP | Senior Representative for Afghanistan and Pakistan |
| USAID | U.S. Agency for International Development |
| USASOC | U.S. Army Special Operations Command |
| VTC | video teleconference |

# CHAPTER ONE

# Introduction

Since late 2001, the United States has been engaged in one of the longest periods of war in its history. Although in historical terms these wars were fought at a relatively low level of lethality, with far fewer casualties than previous major wars, the experiences were frustrating, searing, and somewhat controversial. They have left many Americans wondering if the United States was able to achieve any outcome approximating victory—or at least a satisfactory outcome—in the battlefields of Iraq and Afghanistan, smaller contingencies such as Libya, and in the wider struggle against terrorist groups plotting to attack the United States. While the United States is fortunate in that it does not face an existential threat today comparable to that posed by the Soviet Union during the Cold War, the experience of the past 13 years reveals some troubling lapses, a number of them chronic, that affect the conduct of national security policy and strategy across the spectrum of conflict. An initial study of the lessons from the first decade of war (2001–2011) was produced in June 2012 at the behest of the chairman of the Joint Chiefs of Staff.[1] That study provided a starting point for this RAND Arroyo Center effort, which builds on that work to offer strategic and operational lessons and incorporate insights from the military and interagency policy levels.

While the sponsor of this project is the U.S. Army, and specifically the U.S. Army Special Operations Command, the sponsor agreed that the scope should include the entire joint, interagency, intergov-

---

[1] Joint and Coalition Operational Analysis (JCOA), *Decade of War, Vol. 1: Enduring Lessons from the Past Decade of Operations,* June 15, 2012.

ernmental, and multinational (JIIM) experience, since both Army and all special operations forces (SOF) operate within that wider context. The study sought to identify the most important overarching issues at the levels of policy and strategy and then determine how they affected implementation on the ground. Those lessons are set in the context of the evolving U.S. experience in warfare and are applied to a future operating environment in which irregular and hybrid warfare are expected to play a major if not predominant role. The report then draws implications from this analysis for the joint force, in particular land forces and SOF, operating within a JIIM environment for these types of conflicts.

Chapter Two surveys the U.S. experience in warfare from World War II to the present to identify key trends and inflection points. Warfare became increasingly joint, and technology increased precision in weaponry, improved situational awareness, and enhanced the force's ability to operate at night. SOF became increasingly capable and experienced substantial growth, with historically high rates of operational tempo during the past 13 years. Precision weaponry and refined techniques enabled forces to minimize collateral damage, but at the same time sensitivity to even these lower rates of civilian casualties increased. Armor retained its relevance in lethal environments, and survivability of men and material increased through an array of doctrinal, tactical, and technological improvements. The U.S. military faced ongoing challenges in conducting noncombat missions, such as stabilization, reconstruction, and capacity-building missions—difficulties that became increasingly apparent after 2001. From this, two broad themes emerged. First, land warfare has evolved away from conventional combat against state actors and their standing forces to an increasing incidence of irregular warfare fought by joint forces against nonstate actors. Second, while the joint force has on the whole adapted quickly at the tactical and operational levels, it has often struggled to incorporate the wars' broader strategic lessons.

Chapter Three identifies seven lessons from the policymaking level of government to the operational level of the battlefield, derived from the experiences of 2001–2014. As the chairman's *Decade of War* study noted, "operations during the first half of the decade were often marked by numerous missteps and challenges as the U.S. government and military applied a strategy and force suited for a different threat and environment."[2] This study identifies difficulties in formulating strategy; a tendency to exclude the political dimension from strategy; inadequate understanding of the environment and dynamics of the conflict; repeated failures to prioritize, plan for, and resource stabilization operations, host nation capacity-building, and transitions to civilian control; and inadequate interagency coordination throughout the process from policy formulation through implementation. Over the 13 years, various adaptations permitted partial success, but in many cases those adjustments were belated, incomplete, and ad hoc. Many of these individual shortcomings have been analyzed and debated elsewhere, and some of them are chronic problems that have eluded government attempts at remedies for a variety of reasons. What this study seeks to contribute is a focused set of high-level lessons.

Chapter Four assesses future conflict trends and argues that further adaptation is required to institutionalize the lessons of the past 13 years and prepare for a future that will include frequent irregular and hybrid warfare, according to the National Intelligence Council's Global Trends 2030 projection. The United States faces stark choices about where to invest increasingly scarce defense dollars; retaining an overmatch in conventional military capability is not only necessary to defend against peer competitors and existential threats, but it also forces state and nonstate adversaries to choose a blend of irregular or asymmetric measures to gain advantage. However, the changing character of war necessitates a deeper examination of the basic U.S. approach to war and national security. The *Decade of War* study posited that "the Cold War model that had guided foreign policy for the previous 50 years no longer fit the emerging global environment." If that is true, then a revised theory of success adapted to the current circumstances may provide a compass for strategy. This chapter makes the case for such a theory of success and outlines seven areas for improved JIIM capabilities based on that theory and the lessons derived in Chapter Three.

---

[2]    JCOA, 2012, p. 1.

## Methodology

The approach employed in this study consisted primarily of document-based research and semistructured interviews with experts and officials involved in the decisionmaking and implementation of policies and strategies over the past 13 years. In addition, a workshop of policymakers and academic experts in national security, civil-military relations, and strategy was convened to discuss the policy, strategy, and implementation lessons of the past 13 years.

The RAND Arroyo Center workshop of scholars and policymakers was convened on June 19, 2014, to advance the understanding of how the U.S. government may use all instruments of national power more effectively. Selected academic readings were provided to the participants in advance of the event, and participants were asked to come prepared to articulate three policy- or strategic-level lessons from the past 13 years of war based on their research and experience. This preparation was intended to identify a slate of potential lessons and needed reforms of policy, practice, and/or organization for discussion. The workshop was conducted as a structured discussion to elicit the logical reasoning and experiences of these experts and to debate and prioritize critical issues bearing on the formation and implementation of policy and strategy. The workshop was divided into three discussion modules, each with brief opening remarks by a selected expert followed by three-minute rounds for participant comments.

First, participants discussed the policy process, exploring questions such as the following: Was the process for developing U.S. policy options and defining desired end states effective? Did the policy process include the necessary input and iterative dialogue with the military and intelligence community to inform the development and assessment of options? Were historical examples developed and used correctly?

Second, participants discussed strategy formulation. They debated whether policy goals were effectively matched with available resources to develop a coherent and achievable strategic plan; whether the United States possessed adequate understanding of the environment and the type of conflict it was engaged in; when the United States was slow to adapt its initial strategy, why that was the case; how assessment and strategic adaptation might be made more agile; and whether there is a need to mandate an integrated political-military planning process.

Third, participants discussed implementation. They discussed what changes may be required to improve the U.S. government's conduct of transitions and conflict termination (i.e., Phase IV and V operations); whether the U.S. government has developed the necessary capability to conduct stabilization, reconstruction, and advisory missions; the desired/needed role of civilians in these types of complex contingency operations; and whether the U.S. government has a coherent model for building partner capacity.

The study team then conducted a two-round Delphi poll to identify areas of agreement among the participants. The Delphi technique was originally developed by the RAND Corporation in the 1950s to forecast long-range future trends, events, and outcomes for the military. Employing this technique, this project team developed seven lessons based on the workshop discussion and research. The team then conducted a written poll in which the participants rated the lessons according to their perceived importance and the perceived need for reform or improvement on a five-point scale. The poll was conducted in two rounds. In the second round, the refined list of lessons was rated by a large majority of the participants as important or very important.

Finally, as part of the research on needed capabilities for Chapter Four, the research team identified gaps and shortfalls based on the seven lessons derived in Chapter Three. The team then analyzed the military's adaptations made to date in doctrine, organization, and personnel, as well as civilian interagency and multinational adaptation. To complete its gap analysis, the team examined the work of the Joint Staff irregular warfare executive steering committee to determine whether the relevant practices and capabilities recommended in the 2013 Joint Force Assessment for Irregular Warfare are being institutionalized. A number of that assessment's recommendations are still pending implementation and may be canceled because of budget constraints.

# CHAPTER TWO

# The U.S. Experience in Land Warfare, 1939–2014

On February 25, 2011, Secretary of Defense Robert Gates traveled to the United States Military Academy at West Point for his last speech to the cadets as secretary. In his speech, he reflected that the Army, "more than any other part of America's military, is an institution transformed by war," and that with the wars in Iraq and Afghanistan ending, the Army now faced a new but equally daunting task—ensuring that Iraq and Afghanistan's lessons are not simply "'observed' but truly 'learned'—incorporated into the service's DNA and institutional memory." Learning, Gates continued, does not come easily. The defense establishment often succumbs to what he termed "next-war-itis"—thinking about the future without a proper understanding of the past and appreciation of the demands of present. Tellingly, Gates laced his remarks with historical references drawn not only from the immediate past, but from the last half-century. At the same time, Gates argued, the Army—and, more broadly, the military as a whole—often learns selectively, focusing on what it *wants* to learn and not on what it *needs* to learn. "There has been an overwhelming tendency of our defense bureaucracy to focus on preparing for future high-end conflicts—priorities often based, ironically, on what transpired in the last century—as opposed to the messy fights in Iraq and Afghanistan."[1]

As a result, any study devoted to the lessons of the past dozen years of conflict must first wrestle with Gates' twofold challenge and answer two questions.

---

[1]   Secretary of Defense Robert M. Gates, "Speech to the United States Military Academy (West Point, NY)," February 25, 2011.

- First, how do the wars of the past 13 years fit into the broader evolution of land warfare?
- Second, and more subtly, what lessons did the U.S. military (and, specifically, the Army) learn—and, almost as importantly, not learn—from its wars over the past three-quarters of a century?

This chapter makes two central claims. First, while the Iraq and Afghanistan wars are often portrayed as unique and unprecedented, they actually fit into a broader evolution of land warfare away from conventional combat against state actors to the increasing incidence of irregular warfare fought by joint forces against nonstate actors.[2] Second, as the second half of the twentieth century vividly demonstrates, the Army tends to be adept at learning tactical and operational lessons but less so at learning the strategic lessons.[3] While the Army often learns tactical and operational lessons from the wars it fights, it often struggles to incorporate these war's broader strategic lessons that require it to think outside the box and beyond its immediate past experiences.

## The Formative Years: World War II Through Korea

World War II's influence on the U.S. military, including the Army, cannot be understated. As Hew Strachan argued, "The theoretical force of the Second World War has been with us ever since . . . partly because the conclusion to the war—the dropping of the atom bombs on Hiroshima and Nagasaki—carried its own warnings. As a result total war became the foundation stone for strategic theory in

---

[2]   Irregular warfare (IW) is defined as "[a] violent struggle among states and non-state actors for legitimacy and influence over the relevant population(s). IW favors indirect and asymmetric approaches, though it may employ the full range of military and other capabilities in order to erode an adversary's power, influence, and will." Joint Publication 1, Doctrine of the Armed Forces of the United States, JP I-6, 2013.

[3]   For the broader academic debate about militaries' ability to learn and adapt, see Barry R. Posen, *The Sources of Military Doctrine*, Ithaca, N.Y.: Cornell University Press, 1984; Stephen P. Rosen, "New Ways of War: Understanding Military Innovation," *International Security*, Vol. 13, No. 1, Summer 1988, pp. 134–168.

the second half of the twentieth century."[4] In many respects, World War II revolutionized land warfare, particularly on the tactical and operational levels—as new technologies and concepts were developed, fielded, and tested. The war, however, also produced a certain form of intellectual rigidity, especially on the strategic level, as generations of officers took a largely conventional, firepower intensive, large-army-based view of land warfare. Given World War II's sheer magnitude and geopolitical importance, this bias was understandable, but ultimately, it also proved problematic—as later conflicts increasingly took on an unconventional character.

World War II featured a series of operational breakthroughs in land warfare. In the 1920s and 1930s, the world's better armies drew on World War I's apparent lessons and the ongoing changes in weapons technology to think through how to fight the next war.[5] While no nation formed an entirely optimal ground (much less joint) combat operational concept, the German military developed the most effective approach.[6] Combining a reasonably accurate assessment of the potential of several key new technologies (such as the tank, lightweight voice radios, and higher levels of motor transport in ground forces) with a new concept that stressed seizing the initiative and conducting high-tempo offensive operations (what has become known as *blitzkrieg*, or "lightning war"), the Germans rapidly overwhelmed a number of opponents in the first two years of the war, most notably the Poles and French. Those that survived the initial German onslaught (the British and Soviets) or observed it (the Americans) learned from these tech-

---

[4]   Hew Strachan, *The Direction of War: Contemporary Strategy in Historical Perspective*, New York: Cambridge University Press, 2013, pp. 274.

[5]   For the study of military innovation during the interwar years, see Posen, 1984; Murray Williamson and Allan R. Millett, *Military Innovation in the Interwar Period*, New York: Cambridge University Press, 1998.

[6]   Tim Ripley, *The Wehrmacht: The German Army in World War II, 1939–1945*, Routledge, 2014, pp. 16–18: Richard M. Ogorkiewicz, *Armored Forces, A History of Armored Forces and Their Vehicles,* New York: Arco Publishing, 1970, pp. 20–23, 72–85.

niques and integrated versions of them into their own forces as the war progressed.[7]

Eventually, the United States learned other key lessons as well, including that size matters: Commanders increasingly saw large armies wielding massive amounts of firepower as the key to victory. The Army grew from roughly 190,000 active-duty personnel in 1939 to eight million men and women (to include the Army Air Corps),[8] while the Navy and Marine Corps reached their peak strength of nearly four million personnel by 1945.[9] Major offensives supported by hundreds or even thousands of artillery pieces and aircraft were the norm in World War II. For example, during the breakout from Normandy in late July 1944 (Operation *Cobra*) two American infantry divisions attacked on a front of four miles, supported by the fire of nearly 1,000 division-, corps-, and army-level artillery pieces.[10] Area munitions were used in huge quantities, requiring massive logistics infrastructures to move and distribute such prodigious amounts of ammunition.

As the war progressed, militaries also learned how to better integrate these massive forces. Joint operations—combining land, sea, and naval power—became very important, much more so than was the case in World War I.[11] In the European and Mediterranean theaters, joint operations were mostly air-ground, with an occasional need for ground units to work with naval forces (e.g., the Normandy invasion).[12] In

---

[7]   Jonathan House, *Combined Arms Warfare in the Twentieth Century*, Lawrence, Kan.: University of Kansas, 2001, pp. 64–104.

[8]   National WWII Museum, "By the Numbers: The U.S. Military—U.S. Military Personnel (1939–1945)," undated; HistoryShots, "U.S. Army Divisions in World War II," 2014.

[9]   Navy Department Library, "U.S. Navy Personnel in World War II: Service and Casualty Statistics," *Annual Report, Navy and Marine Corps Military Personnel Statistics*, June 30, 1964.

[10]   Max Hastings, *Overlord, D-Day and the Battle for Normandy*, New York: Simon and Schuster, 1984, pp. 250–251.

[11]   Jeter A. Isely and Philip A. Crowl, *The Marines and Amphibious War*, Princeton, N.J.: Princeton University Press, 1951, pp. 70–78.

[12]   Murray Williamson and Allan R. Millett, A *War to Be Won: Fighting the Second World War*, Cambridge, Mass.: Harvard University Press, 2009; Max Hastings, *Overlord, D-Day and the Battle for Normandy*, New York: Simon and Schuster, 1984, pp. 250–251.

the Pacific, ground forces operated much more frequently with navies because of the geography of the region.

As innovative as war was on the tactical and operational levels, the World War II experience also produced a series of intellectual blinders when it came to thinking about the future of warfare. To begin, the war ingrained a conventional bias in the United States military. From the American military's perspective, World War II was overwhelmingly a conventional conflict. Only those few Americans who avoided capture in the 1941–1942 Philippine campaign and joined Filipino guerilla groups to harass the Japanese occupiers and the relatively few Americans in the Office of Strategic Services (OSS) gained meaningful experience in guerilla war. Of the roughly eight million serving with the United States Army by the end of the war, fewer than 24,000 served with the OSS.[13]

Second, and somewhat surprisingly, the World War II experience also deemphasized the importance of reconstruction after conflict. Similar to the experience at the end of the First World War when the military had to provide civil administration and services until new local governments were established, after World War II the Allies occupied and administered Axis and liberated territory, including West Germany (1945–1955), Japan (1945–1952), Austria (1945–1955), Italy (1943–1946), and South Korea (1945–1948).[14] These were the U.S. military's largest and most extensive experiences in what today is called stability operations. In Germany they were initially led by Gen. Lucius Clay, and after 1949 by John J. McCloy, while the military led the occupation of postwar Japan. Notably, the occupations lasted longer than the war itself, suggesting just how important the Army's role was in these nonkinetic operations even in this era of large-scale, indus-

---

[13]   The National Archives states that there are 23,973 personnel files from the OSS between 1941 and 1945. Although small in size, the OSS led to the subsequent development of the Special Forces, its unconventional warfare doctrine, and the Army's psychological warfare department. National Archives, "Organization of the Office of Strategic Services (Record Group 226)," undated.

[14]   U.S. Department of State, Office of the Historian, "Occupation and Reconstruction of Japan, 1945–52," undated.

trial warfare.[15] However, in the downsizing of the U.S. military after World War II, the postwar governance operations were not recognized in military training and doctrine as core military missions. Instead, they were seen as one-off events—despite the military's involvement in various small peace operations during the Cold War, such as in Lebanon in 1958 and the Dominican Republic in 1965.

Third, and for understandable reasons, the war focused attention on nuclear weapons and fighting on the nuclear battlefield. The atomic bomb attacks on Hiroshima and Nagasaki in Japan ushered in a new nuclear age, and this new technology would have a profound influence on future military operations. In the coming decades, the Army developed an array of tactical nuclear weapons and developed doctrine for operating on a battlefield where tactical nuclear weapons might be used. While this was a logical, if prudent, choice within the context of the Cold War, ultimately, it produced strategies and an overarching mindset within the Army that—in the eyes of many analysts—were ultimately ill-adapted to wars the United States actually would fight in the second half of the century.[16]

In many ways, the Korean War (1950–1953) seemed to confirm many of World War II's lessons and further ingrained the conventional perspective within the Army. The weapons and tactics used in Korea were essentially the same as in World War II. And like in World War II, this was largely a conventional fight.[17] In an effort to compensate for Communist numerical superiority and to hold friendly casualties to a reasonable level in what was an increasingly unpopular war, the Army employed firepower on a massive scale, like it had in World War II.[18]

---

[15] The post–World War II occupations were different in both scope and scale from the Allied occupation of the Rhineland at the end of the First World War. The occupation of the Rhineland was a military measure designed to create a buffer between France and Germany; it did not aim at the political reconstruction and democratization of Germany.

[16] Andrew F. Krepinevich, *The Army and Vietnam*, Baltimore: Johns Hopkins University Press, 1986.

[17] Jonathan House, *Combined Arms Warfare in the Twentieth Century*, Lawrence, Kan.: University of Kansas, 2001, pp. 196–205. Special operations and CIA operations were also conducted in the Korean War.

[18] Bruce Cumings, *A Korean War: A History*, New York: Random House, 2010.

By late 1951 the fighting in Korea bogged down into World War I–like trench combat, with United Nations (mostly American) forces fighting regimental- and division-sized battles over hilltops of little tactical value. Ultimately, the Korean War fought to a bloody draw.[19] The less-than-successful outcome of the Korean War, however, did not fundamentally shake many of the Army's basic assumptions about the conventional, firepower-intensive, large-army nature of land warfare. Instead, by the mid-1950s there was considerable focus on the defense of Europe against possible Soviet aggression. Considerable amounts of thought went into the issue of how ground combat would take place in Europe in a conflict that included the possible use of nuclear weapons.

Ultimately, the period from World War II to the Korean War highlights the ability of the Army to learn a variety of tactical and operational lessons. Over the course of both conflicts, it assimilated new technologies and new tactics to fight conventional wars better. It proved less adept at thinking outside of the box and imagining alternatives to the World War II model of land warfare. Unfortunately for the Army, the World War II model of warfare was quickly becoming outmoded.

## The Model Breaks Down: The Vietnam War Through the 1980s

While much of the attention remained focused on deterring the Soviets from invading Europe, the United States also faced a new threat in the post–Second World War era—the rise of Communist insurgencies. As the Kennedy administration entered office in 1961, many parts of the Third World were convulsed by the increasingly messy process of decolonization and the "wars of national liberation."[20] The administra-

---

[19] Max Hastings, *The Korean War*, New York: Simon and Schuster, 1987, pp. 333–335.

[20] Colonial powers were gaining ample experience in many of these struggles, which did not involve U.S. forces. Insurgent and irregular warfare occurred in the Arab Revolt in Palestine (1936–1939), the Zionist campaign against the British, the Malayan Emergency (1948–

tion realized that should the communists elect to undermine U.S. and Western interests via guerilla warfare, insurgency, and revolutions, the U.S. nuclear arsenal would be irrelevant. And so, the military needed to find alternative ways of combating this type of threat. For its part, the Army's ability to adapt to this new threat proved uneven—as highlighted by its searing experience during the Vietnam War.

On the tactical and operational levels, the Army in Vietnam showed some signs of innovation and embraced new technologies relatively quickly. Ground combat in South Vietnam included the first widespread use of helicopters in warfare. Initially developed during World War II and refined in the 1950s, helicopters were used for troop transport, medical evacuation, and fire support (gunship) in Vietnam. Showing great promise in the middle years of the war, by the end of the conflict (1971–1972 for U.S. forces) the helicopter was increasingly vulnerable because of the introduction of shoulder-fired surface-to-air missiles and better quality anti-aircraft guns. Approximately 5,000 helicopters were lost during the Vietnam War, roughly half due to enemy action.[21]

Similarly, by the end of the war, the U.S. military also introduced other technologies that would later define modern combat. By 1972, the last year of the war, the Army was using wire-guided anti-tank missiles launched from helicopters, marking the beginning of a decades-long shift toward precision weaponry.[22] Likewise, beginning in the 1960s and accelerating rapidly after the conflict, the Army began to field night vision systems, marking the beginning of what would become one of the American military's major tactical advantages over many of its opponents in the subsequent decades.[23]

---

1960), the Algerian War of Independence (1954–1962), and the Cypriot War of Independence (1955–1960), among many others.

[21] Simon Dunstan, *Vietnam Choppers, Helicopters in Battle 1950–1975*, London: Osprey Publishing, 2003, pp. 200–201.

[22] Dunstan, 2003, pp. 94–106.

[23] Thomas Mahnken, *Technology and the American Way of War Since 1945*, New York: Columbia University Press, 2010.

Where the Army struggled to adapt, however, was more on the conceptual level. During the war, the United States faced a twofold threat—a conventional threat from the North Vietnamese Army and an insurgency in the form of the Viet Cong. Although the Army had faced nonstate actors in the post–World War II landscape before (such as with the Hukbalahap Rebellion in the Philippines), none matched the Viet Cong's size and sophistication.[24] Nevertheless, the Army found itself at a strategic loss on how to respond to this form of warfare.[25] Having largely ignored the unconventional side of World War II and the smaller earlier wars of the post–World War II era, the Army of the early to mid-1960s had little doctrine or experience in irregular warfare. More often than not, the Army turned to its roots in the tactics of World War II and Korea to find conventional solutions. Just as in World War II, the military expended huge quantities of artillery and air-delivered ordnance in South Vietnam, Laos, and North Vietnam. Similarly, Army units were organized and fought much along World War II norms, with the notable addition of the helicopter capability.[26]

As the war progressed, ground forces gradually improved their understanding of irregular warfare. Some tactics optimized for counterinsurgency were introduced, and more effort was devoted to improving the South Vietnamese forces. The 5th Special Forces Group, for example, organized and trained Civilian Irregular Defense Groups (CIDG) in the mountainous border regions of Vietnam. Eventually amounting to a 50,000-man army, mostly recruited from ethnic minorities, the

---

[24] The Hukbalahap Rebellion was from 1949 to 1951 and ranged in size from 11,000 to 15,000 actors: Benedict J. Kervkliet, *The Huk Rebellion: A Study of Peasant Revolt in the Philippines,* University of California Press, 1977, p. 210. By comparison, the numbers for the People's Army of Vietnam range from 240,000 in 1960 to 643,000 in 1975; see Correlates of War data set: Max Boot, *The Savage Wars of Peace: Small Wars and the Rise of American Power,* New York: Basic Books, 2002, pp. 281–335.

[25] For the debate over the Army's lack of strategy in Vietnam, see Andrew F. Krepinevich, *The Army and Vietnam,* Baltimore: Johns Hopkins University Press, 1986; Harry Summers, *On Strategy: A Critical Analysis of the Vietnam War,* Novato, Calif.: Presido Press, 1995.

[26] Krepinevich, 1986, pp. 3–26.

CIDG collected intelligence and helped secure this difficult terrain.[27] More broadly, in 1967, the Johnson administration created the Civil Operations and Revolutionary Development Support (CORDS) program—an interagency effort that sought to coordinate the military, the Department of State, the U.S. Agency for International Development, and the Central Intelligence Agency's efforts to "pacify" the Vietnamese countryside. Indeed, some attribute the decline of the Viet Cong insurgency in South Vietnam during the second half of the war to the CORDS program's success.[28] Despite bright spots like the CORDS program, however, most of the Army did not adapt to the challenges of counterinsurgency; the primary orientation of the U.S. ground forces focused on conventional combat from the start of the war to the end.[29]

Perhaps the most-cited example of the Army failing to learn from Vietnam's big lessons occurred after the war was over. Once U.S. forces withdrew, the Army moved away from preparing for irregular warfare, with the notable exception of the Special Forces.[30] The Special Forces, however, shrank to a fraction of its wartime size. The Marine Corps retained more of its counterinsurgency capability, but it also refocused mostly toward conventional operations.[31] Army historian Conrad Crane described the post-Vietnam institutional attitude as fol-

---

[27]  *Vietnam Studies: U.S. Army Special Forces, 1961–1971*, Center for Military History Publication 90-23, Washington, D.C.: Department of the Army, 1989; Robert M. Cassidy, "Back to the Street Without Joy: Counterinsurgency Lessons from Vietnam and Other Small Wars," *Parameters*, Summer 2004, p. 76.

[28]  CORDS and future CIA director William Colby even goes far as to label the period 1971–1972 as "victory won" and suggest that thanks to CORDS the 1972 offensive included "no substantial guerrilla action." William Colby and James McCargar, *Lost Victory: A First-hand Account of America's Sixteen-Year Involvement in Vietnam*, Chicago, Ill.: Contemporary Books, 1989, pp. 291, 363.

[29]  Krepinevich, 1986; John Nagl, *Learning to Eat Soup with a Knife: Counterinsurgency Lessons from Malaya and Vietnam*, 2002.

[30]  Krepinevich, 1986, pp. 268–274. Note, however, that the Army's particular focus during this period was the existential threat posed by the Soviets in Europe. While this had to be the main effort, it may have been more beneficial for the Army to retain more of the lessons from Vietnam.

[31]  In fact, in the 1980s, some considered devolving these "small wars" (the earlier term for unconventional conflicts) to the Marine Corps, while the Army would focus on conven-

lows: "The U.S. Army's assessment of its failure in Vietnam was quite different from the French. . . . Army involvement in counterinsurgency was first seen as an aberration and then as a mistake to be avoided."[32]

While the U.S. military may have turned its back on irregular warfare, the world did not. In the decade after Vietnam, numerous "small wars" raged in many parts of the world. Central America, Colombia, South and Southeast Asia, and much of sub-Saharan Africa played host to protracted insurgencies that left thousands dead and occasionally changed the political balance in a region. The Soviets overthrew the Afghan government in December 1979 in a seemingly highly successful, rapid *coup de main*, only to become bogged down in a decade-long insurgency that left over 15,000 Soviet soldiers and several hundred thousand Afghans dead. While U.S. covert action occurred in Afghanistan, Angola, and Nicaragua, most of the conventional Army prepared for conventional combat with the Soviets in Europe. Contingency operations in Grenada and Panama in 1983 and 1989 showed that the U.S. military was poorly prepared for joint operations, much less to conduct stability operations upon the toppling of the military regime.[33]

Ultimately, Vietnam marked a pivot point for American land warfare. On the one hand, it represented the end of the era of Industrial Age warfare. As the war came to a close, conscription (in use since 1940) ended, and the military converted to being an all-volunteer force. Also, the Vietnam War was generally fought with the same industrial warfare concepts that had been the norm since World War II and earlier. Huge quantities of mass-produced, largely unguided weapons were employed in conjunction with large ground forces (peak U.S. strength in Vietnam by early 1969 was roughly 536,000 personnel).[34] At the same

tional, large-scale warfare. See Eliot A. Cohen, "Constraints on America's Conduct of Small Wars," *International Security*, Vol. 9, No. 2, Fall 1984, pp. 178–179.

[32] Conrad Crane, *Avoiding Vietnam, the U.S. Army's Response to Defeat in Southeast Asia*, U.S. Army War College, 2002, p. 2.

[33] R. Cody Phillips, *Operation Just Cause: The Incursion into Panama*, Washington, D.C.: Center of Military History, 2006, pp. 42–43.

[34] Bob Seals, "The 'Green Beret Affair': A Brief Introduction," Military History Online, 2007.

time, Vietnam also marked the first major attempt to think through the challenges of irregular warfare—with mixed success. While the Army proved readily able to adapt to new technologies—like helicopters or night vision equipment—it proved less capable of understanding and embracing the greater strategic shift under way from an era marked by conventional warfare to one in which it would find itself increasingly engaged in irregular warfare.

## The 1990s and the Search for a New Paradigm

The Soviet Union's sudden collapse between 1989 and 1991 upended the strategic assumptions that had been in place for almost half a century. The United States and its Western allies no longer needed to worry about the Red Army's tanks pouring through the Fulda Gap. Europe seemed—at least for the moment—safe and secure. As a result, during the 1990s, all the NATO militaries were significantly reduced. For land forces in particular, the collapse of the Soviet Union meant a search for a new paradigm of warfare: What sort of threats would the United States face now that the Soviets were gone, and what would the land component's role be in this new unipolar world? The answers to both questions were far from clear; indeed, the Army experimented with three different models for the future of warfare during the 1990s.

Initially, the future of land warfare seemed to be very much in keeping with the past: The first Persian Gulf War (Desert Storm) saw a major ground operation based largely on World War II operational concepts. Following its seizure of Kuwait in August 1990 in an armored assault, Saddam Hussein's army dug into southern Iraq and Kuwait and passively watched the buildup of a massive U.S.-led coalition force in Saudi Arabia. From August 1990 to February 1991, over 530,000 U.S. military personnel (of whom nearly 300,000 were Army) deployed to the region to prepare for a counteroffensive to retake Kuwait. Additionally, there were several hundred thousand coalition soldiers, sailors, and airmen.[35]

Operation Desert Storm started with 38 days of intense air attacks followed by a ground offensive that included 17 division-sized formations (seven U.S. Army, two Marine Corps, two Egyptian, one Syrian, two Saudi, one Gulf States, one French, and one British), plus several independent regimental-sized units. The result was a rapid, overwhelming victory over the Iraqis, with minimal coalition losses. Thanks to precision weapons and relatively open desert terrain, air attacks reduced many Iraqi units deployed in fixed defenses along the Saudi-Kuwaiti border to strengths of 50 percent or less.[36] During the few significant tactical ground battles, the U.S. M-1 Abrams and the British Challenger tanks achieved lopsided victories against those Iraqi conventional units that attempted to stand and fight. Extremely effective armor protection, excellent night fighting capabilities, and superior training made the company- and battalion-sized engagements easy wins for the coalition.[37] Army Apache attack helicopters developed for use against Soviet armor in Europe also performed well in this conflict. Even some of the less-than-effective operations—like the SOF's hunt for Iraqi SCUD mobile missile launchers—could claim that they had helped set the strategic conditions for a decisive coalition victory.[38]

---

[35] Frank Schuber et al., *The Whirlwind War*, Washington, D.C.: Center of Military History, 1995, p. 157

[36] Of the roughly 88,000 tons of aerial munitions that were expended during the 42 days of combat, roughly 8 percent were guided munitions. "Emergence of Smart Bombs," *Air Force Magazine*, 2014.

[37] Stephen T. Hosmer, *Psychological Effects of U.S. Air Operations in Four Wars, 1941–1991: Lessons for U.S. Commanders*, Santa Monica, Calif.: RAND Corporation, MR-576-AF, 1996, pp. 43–68. Apaches were also credited with killing several hundred Iraqi vehicles at minimal cost.

[38] William Rosenau, for example, argues that although SOF may not have destroyed as many SCUD launchers as intended, they succeeded at keeping Israel out of Desert Storm (by telling the Israelis that the coalition's "best trained, most experienced, and most elite ground forces" were on the mission), mitigating what could have otherwise been a messy political situation. William Rosenau, *Special Operations Forces and Elusive Enemy Ground Targets: Lessons from Vietnam and the Persian Gulf War*, Santa Monica, Calif.: RAND Corporation, MR-1408-AF, 2001, pp. 43–44.

The U.S. military, understandably, was proud of its success in Desert Storm: This was, after all, the war the Army had prepared to fight ever since World War II. There was no messy irregular warfare component or a postwar stabilization mission (although the Iraqi regime's attacks on its Kurdish population prompted Operation Provide Comfort, in which conventional forces and SOF provided relief to the Kurds). Desert Storm was the war the Army had spent the post-Vietnam period training for, just fought on easier—or at least more open—terrain and against a far less formidable adversary.[39] On a deeper level, the war fit with many of the Army's deep-seated notions about how wars *should* be fought: The plan of attack and the massive size of the force would have been entirely familiar to a World War II–era commander. The Army, in short, was well within its strategic comfort zone.

After Desert Storm, the United States participated in a number of small military operations that did not involve large-scale conventional combat. First called "peacekeeping" missions but later renamed "peace-building," "postconflict reconstruction and stabilization," or "stability operations," the American ground forces deployed to Somalia (1992–1994), Haiti (1994–1995), Bosnia (1995–present), and Kosovo (1999–present). Of these campaigns, the Army, however reluctantly, gained the most experience with stability operations in the Balkans. Eventually, the Balkans provided an alternative use for land forces in the post–Cold War era.

The Army first deployed to Bosnia-Herzegovina in Operation Joint Endeavor following the 1995 Dayton Accords as part of a NATO-led, 36-nation, 60,000-strong international force (IFOR). About 18,000 personnel, primarily from the 1st Armored Division, formed the core of Multinational Division (North), and another 10,000 U.S. personnel served as part of various NATO and U.S. headquarters and support elements.[40] IFOR's primary goals were to establish a sustainable cessation of hostilities, ensure force protection, and estab-

---

[39] Crane, 2002, pp. 15–18.

[40] Larry Wentz, *Lessons from Bosnia: The IFOR Experience*, Washington, D.C.: DoD CCRP/ NDU, 1997, pp. 3–4.

lish enduring security and arms control measures.[41] IFOR enforced the zone of separation between the former belligerents, monitored the withdrawal of heavy weapons, assisted international organizations in their humanitarian missions, and observed and prevented interference in the movement of civilian populations, refugees, and displaced persons.[42] In a forerunner to future high-value-targeting missions in the next decade, American SOF also embarked on a campaign to capture war criminals.[43]

At the end of the 1990s, U.S. forces again joined a NATO coalition, this time in Kosovo. NATO first attempted an air campaign to coerce the Serbian government to cease its aggressive actions against the largely ethnic Albanian population in the province of Kosovo. While NATO deployed ground troops—including the Army's Task Force Hawk—senior NATO civilian and military leaders were reluctant to plan, much less execute, a ground attack into Kosovo, which to some extent ceded the initiative to Belgrade.[44] In the end, a ground assault was not needed, and 78 days after hostilities began, Belgrade acquiesced to NATO's demands. The United States, however, did send 7,000 troops as part of a 50,000-man NATO force to assist in stability operations and pave the way for a new government there,[45] and as of 2014, some 940 U.S. troops still remain in the Balkans as part of a 5,000-man peace support force.[46]

---

[41] James Dobbins, John G. McGinn, Keith Crane, Seth G. Jones, Rollie Lal, Andrew Rathmell, Rachel Swanger, and Anga Timilsina, *America's Role in Nation-Building: From Germany to Iraq*, Santa Monica, Calif.: RAND Corporation, MR-1753-RC, 2003, p. 93.

[42] *General Framework Agreement for Peace in Bosnia and Herzegovina*, a.k.a. "Dayton Accord," 1995, Annex IA.

[43] Charles Cleveland, "Command and Control of the Joint Commission Observer Program U.S. Army Special Forces in Bosnia," Strategy Research Project, United States Army War College, 2001.

[44] Bruce R. Nardulli, Walter L. Perry, Bruce Pirnie, John Gordon IV, and John G. McGinn, *Disjointed War, Military Operations in Kosovo, 1999*, Santa Monica, Calif.: RAND Corporation, MR-1406-A, 2002, p. 111.

[45] Nardulli et al., 2002, Chapter Five.

[46] Army Posture Statement, Testimony Before Senate Armed Services Committee, April 3, 2014.

Adapting to this second paradigm of land warfare—in which ground forces became peacekeepers, politicians, and development specialists—proved more difficult. Assessments and after-action interviews found that U.S. forces were not adequately prepared to operate as part of multinational coalitions. Their training still focused almost exclusively on warfighting, not on the wide range of nontraditional challenges of stability operations. In these situations, like the Balkans, they needed to include how to negotiate with factional leaders and local offices, manage civil-military relations, and ensure a safe and secure environment for implementation of the peace accords.[47] Similarly, integrating the special operations community into this environment proved equally challenging: "conventional U.S. commanders in Bosnia, initially skeptical, and even hostile, toward SOF (special operations forces) operations."[48] Eventually the relationship between the conventional and special operations communities warmed and laid the groundwork for even closer coordination during the next decade of war, but this process took time.[49]

The Army also experimented with yet a third model for the future of land warfare during the 1990s.[50] Beginning in the middle of the decade, some strategists argued that warfare was in the midst of a "revolution in military affairs (RMA) and was shifting to a new era of network-centric warfare (NCW). They argued that a combination of (1) greatly improved sensor capabilities, (2) increasingly sophisticated command and control systems, and (3) precision weapons allowed the United States to accurately strike targets from afar, with relatively

---

[47] Dobbins et al., 2009; and Howard Olsen and John Davis, "Training U.S. Army Officers for Peace Operations: Lessons from Bosnia," United States Institute of Peace Special Report, October 29, 1999.

[48] Matthew Johnson, "The Growing Relevance of Special Operations Forces in U.S. Military Strategy," *Comparative Strategy*, Vol. 25, No. 4, 2006, p. 283.

[49] Johnson, 2006, p. 283.

[50] Paul K. Davis, *Military Transformation? Which Transformation, and What Lies Ahead?* Santa Monica, Calif.: RAND Corporation, RP-1413, 2010, pp. 11–14.

little (if any) need to commit ground forces.[51] According to this view, the United States could replace the massive armies wielding immense firepower, like those it employed in World War II, Vietnam, or even Desert Storm, with investments in intelligence collection and precision strike capabilities.

Throughout the mid- to late 1990s and for some time after the September 11, 2001, attacks, the NCW concept was gaining acceptability, particularly among airpower advocates. Indeed, the Kosovo air war seemed to confirm the idea that wars could be won with minimal ground forces.[52] Even within the Army, there was a considerable amount of acceptance of these concepts, particularly the claims that much of the "fog of war" would be swept away by the powerful combination of advanced sensors and their associated processing capabilities. Ultimately, it took the experience of combat in Iraq and Afghanistan from 2002 to 2014 to seriously question the viability of these ideas.

Over the 1990s, then, the Army found itself in a period of strategic flux. Without the Red Army, it became increasingly unlikely that the future of land warfare would consist of massive armies locked in large-scale, conventional combat. Indeed, as Desert Storm vividly demonstrated, any foe choosing to engage the United States military in conventional combat was risking disaster. That said, if conventional combat was not the future of land warfare, it was not clear what would take its place. The Army struggled to learn from its role in stability operations in the Balkans and elsewhere, all while it was trying to find its place in the yet-untested concepts of NCW. For the Army, it was just the first steps in learning to think "big."

---

[51]  Clay Wilson, "Network Centric Warfare, Background and Oversight Issues for Congress," Congressional Research Service, June 2, 2004.

[52]  Daniel L. Byman and Matthew C. Waxman, "Kosovo and the Great Air Power Debate," *International Security,* Vol. 24, No. 4, Spring 2000, pp. 5–38.

## From NCW to Counterinsurgency: Afghanistan and Iraq, 2001–2014

The initial military response to the September 11, 2001, attacks seemed to conform to the NCW's assertions about the future of warfare. The United States deployed only a few hundred CIA and military SOF personnel to Afghanistan in late 2001 and early 2002, supported by Air Force bombers and Navy fighters, to partner with large numbers of Afghan fighters. The "network" seemed to work well: The few U.S. personnel on the ground passed targeting information to aircraft to strike Taliban forces, while Afghan troops, primarily members of the Northern Alliance, comprised most of the ground force. By early 2002, the Taliban regime had been toppled. The victory, however, was not perfect. The absence of U.S. ground forces probably permitted Osama bin Laden and the senior al Qaeda leadership to escape across the border into Pakistan.[53] Despite this failing, Afghanistan's apparent success directly influenced U.S. strategic thinking about the next campaign: to depose Saddam Hussein in Iraq.

The March–April 2003 march to Baghdad was accomplished by an air-ground force that was a fraction of the size of the 1991 Desert Storm operation.[54] Those Iraqi military units that attempted to resist were easily routed, and the poorly trained Iraqi militia elements were

---

[53] There is some debate about whether more American ground forces would have prevented bin Laden's escape into Pakistan. In his analysis in *Foreign Affairs*, Stephen Biddle argued, "at Tora Bora, massive American bombing proved insufficient to compensate for allied Afghan unwillingness to close with dug-in al Qaeda defenders in the cave complexes of the White Mountains. This ground force hesitancy probably allowed bin Laden and his lieutenants to escape into neighboring Pakistan" (Stephen Biddle, "Afghanistan and the Future of Warfare," *Foreign Affairs*, Vol. 82, No. 2, March–April 2003, p. 43). Similarly, Brookings scholar Michael O'Hanlon concluded, "the prospects for success (in capturing Bin Ladin [sic]) in this case were reduced considerably by U.S. reliance on Pakistani forces and Afghan militias for sealing off enemy escape routes and conducting cave-to-cave searches during critical periods" (Michael O'Hanlon, "Flawed Masterpiece," *Foreign Affairs*, Vol. 81, No. 3, May–June 2002, p. 48). As with any historical counterfactual, it is impossible to prove definitively.

[54] Amy Belasco, *Troop Levels in Afghan and Iraq Wars, FY2001–FY2012: Cost and Other Potential Issues,* Congressional Research Service, Government Printing Office, July 2, 2009, p. 35. According to this chart, the average strength for FY 2003 was 216,500 military personnel.

even easier to defeat. But while the operation was conducted with relative ease, there were some warning signs that the promises of NCW RMA might not be realized. The lower-level tactical ground units (division and below) had poor situational awareness on the advance to Baghdad. Indeed, the tactical intelligence system largely collapsed, leaving the advancing ground forces with little better than World War II levels of awareness of the enemy. The effectiveness of U.S. and British heavy armor compensated for the very poor quality of tactical situational awareness.[55]

By late 2003, as the Iraq and Afghanistan wars moved into their counterinsurgency phases, it became apparent that NCW could not provide the degree and type of battlefield awareness that was needed. With the enemy able to blend into the local populations, locating insurgents proved difficult, and NCW sensors, originally designed to detect conventional enemy troops and equipment, could not provide the intelligence that commanders needed. Similarly, once improvised explosive devices (IEDs) became the insurgents' weapon of choice, Coalition troops found that they needed both better ways of detecting these roadside bombs and more armored vehicles to defend against these attacks—rather than the light, highly mobile forces NCW predicted. Indeed, by mid-decade the vehicles employed in Iraq and Afghanistan became much heavier because of the desire to add armor protection. Above all, Iraq and Afghanistan demonstrated that NCW's premise that technology could substitute for ground troops went only so far. Indeed, ground forces proved essential to establish some degree of protection for and control of the population and to engage the insurgents in Afghanistan and Iraq.

NCW's failure in Iraq and Afghanistan spurred a twofold response in the Army. On the technical level, as in Vietnam, the Army turned to technology to provide the answer to the challenges of irregular warfare.

---

[55] John Gordon and Bruce Pirnie, "Everybody Wanted Tanks, Heavy Forces in Operation Iraqi Freedom," *Joint Force Quarterly*, Issue 39, 2005, pp. 84–90. Two bright spots in initial stages of Operation Iraqi Freedom were the very effective, timely air support that the ground forces received, as well as the much better situational awareness of the location of friendly forces due to the just-fielded Blue Force Tracker system that provided near-real-time updates of the whereabouts of friendly (Blue) units.

Beginning in 2003, the Army Chief of Staff set up the Army IED Task Force, later the Joint IED Defeat Organization.[56] The organization pioneered technical advances in order to detect and mitigate the threat of IEDs. Eventually, the Army harnessed advances in biometric, imagery, and signals intelligence to help unmask some of the insurgent threat, and it fielded vehicles with V-shaped hulls that proved more survivable against IED blasts. On the tactical end, Army units—led by the special operations community—developed the fusion of operations and intelligence for the purpose of hunting high-value targets into a high art, facilitated by greatly expanded intelligence collection and advances in information technology.

Perhaps the more challenging transformation for the Army, however, was on the strategic level—shifting ground forces away from NCW to a counterinsurgency (COIN) paradigm. Similar to the Vietnam experience, the U.S. forces initially deployed to Iraq and Afghanistan had focused on large-scale maneuver warfare or else on the peacekeeping missions of the Balkans, and so were ill-prepared to fight a COIN. Indeed, the Army found itself relearning many of the lessons from Vietnam in Afghanistan and Iraq.[57] To further complicate matters, the relatively small size of the Army during Iraq and Afghanistan compared to the previous conflicts in the period left the Army in need of skills necessary for COIN, including critical enablers like military police, civil affairs, and explosive ordnance disposal.[58]

Only after years of struggling did the Army gradually adapt to the COIN mission. In 2006, the Army and Marine Corps produced *FM 3-24: The Counterinsurgency Manual*, one of the first concerted

---

[56] Joint IED Defeat Organization, "About JIEDDO," undated.

[57] For an example, see Nagl, 2002.

[58] For the size of the force, see Amy Belasco, "Troop Levels in the Afghan and Iraq Wars, FY2001–FY2012: Cost and Other Potential Issues," Washington, D.C.: Congressional Budget Office, 2009. For the importance of enablers to reconstruction, see Conrad C. Crane, "Phase IV Operations: Where Wars Are Really Won," *Military Review*, May–June 2005, pp. 27–36.

attempts to produce a comprehensive COIN approach.[59] The manual argued that many of the central tenets of the large-army, firepower-intensive World War II way of land warfare were precisely what forces should avoid doing COIN. Instead, it argued, COIN required ground forces to focus on developing the economy, promoting local government, providing essential services, and training at least as much as on more kinetic operations. While the actual impact of many of the manual's prescriptions on Iraq and Afghanistan operations remains hotly contested, there is no denying its influence.[60] Perhaps more significantly, from an institutional perspective, it took several years—and the Army facing a "fiasco" in Iraq, as one prominent journalist titled his book at time—for the Army to "think big" and break from its conventional mindset.[61]

The other notable evolution was the enormous increase in the use of SOF in a wider array of missions than at any other time in their history. Their intensive use in Afghanistan, Iraq, and other venues of the Global War on Terrorism, as it was called for much of the decade, began an unfinished journey in breaking the conventional mindset. While much of the attention focused on Iraq and Afghanistan, the war on terrorism was also prosecuted in regions stretching from the Philippines to Yemen. Much of the burden fell on SOF: Indeed, in the period from 2001 to 2014, SOF experienced the highest operational tempo in their history and more than doubled in size to a total of some 33,000 uniformed operators, about half of whom were Army soldiers,

---

[59] For evolution of the Counterinsurgency Manual and its impact on strategy, see Fred Kaplan, *The Insurgents: David Petraeus and the Plot to Change the American Way of War*, New York: Simon & Schuster, 2013; and Linda Robinson, *Tell Me How This Ends: General David Petraeus and the Search for a Way Out of Iraq*, New York: PublicAffairs, 2008.

[60] For some of the intense debate around the manual's effectiveness, see John A. Nagl, "Constructing the Legacy of Field Manual 3-24," *Joint Forces Quarterly*, Vol. 58, No. 3, 2010, pp. 118–120; Gian P. Gentile, "Time for the Deconstruction of Field Manual 3–24," *Joint Force Quarterly*, Vol. 58, No. 3, 2010, pp. 116–117; Stephen Biddle, Jeffrey A. Friedman, and Jacob N. Shapiro, "Testing the Surge: Why Did Violence Decline in Iraq in 2007?" *International Security*, Vol. 37, No. 1, Summer 2012, pp. 7–40; and Raphael S. Cohen, "A Tale of Two Manuals," *Prism*, Vol. 2, No. 1, 2010, pp. 87–100.

[61] Thomas E. Ricks, *Fiasco: The American Military Adventure in Iraq*, New York: Penguin Press, 2006.

and over 69,000 total personnel assigned to the U.S. Special Operations Command.[62] In Afghanistan, SOF did not just accomplish the initial toppling of the Taliban in conjunction with airpower, the CIA, and Afghan militias; thousands of operators remained throughout the war distributed in small-fire bases to work with various Afghan regular and irregular forces in counterterrorism, capacity-building, and local defense activities.[63] In Iraq, SOF were similarly engaged in a wide range of missions throughout the war, starting with their distribution in the west, south, and north, where operators allied with Kurdish militias pinned down 11 divisions of the Iraqi army.[64] Subsequently, some 5,000 operators remained to build and conduct operations with Iraqi special units, engage with tribes, and prosecute unilateral counterterrorism missions.

The expansion in demand for SOF prompted a variety of tactical innovations, especially in terms of intelligence sharing and targeting, allowing these forces to find, fix, and finish terrorist targets at ever more rapid rates.[65] These tactics were migrated from the major theaters of war to Yemen, Africa, and elsewhere; in addition, SOF began less-well-documented but long-term activities to support indigenous forces battling insurgency and terrorism in the Philippines, Colombia, East Africa, and West Africa. On a more profound level, more structural innovations began to break down traditional boundaries between conventional and SOF, and between the military, the intelligence community, and law enforcement personnel. SOF operated in close proximity to and in support of conventional forces throughout the wars in Afghanistan and Iraq, and in both cases conventional maneuver

---

[62] William H. McRaven, "Testimony to the U.S. Senate on the Department of Defense Authorization of Appropriations for Fiscal Years 2015 and the Future Years Defense Program," U.S. Senate Armed Services Committee, March 11, 2014.

[63] Charles H. Briscoe et al., *Weapon of Choice: ARSOF in Afghanistan*, Fort Leavenworth, Kan.: Combat Studies Institute Press, 2003; Linda Robinson, *One Hundred Victories: Special Ops and the Future of American Warfare*, New York: PublicAffairs, 2013.

[64] Charles H. Briscoe et al., *All Roads Lead to Baghdad: Army Special Operations Forces in Iraq*, New York: Paladin Press, 2007.

[65] Stanley A. McChrystal, *My Share of the Task: A Memoir*, New York: Portfolio/Penguin, 2013.

units were assigned to support high-value targeting and local defense missions. Despite this increasing interaction and the efforts of special operations leaders to develop new doctrine, operational concepts, and command structures, the evolution of special operations as a tool capable of strategic impact and the codification of its integrated application with the rest of the joint force and other elements of power remained incomplete.

## Learning the Big Lessons

This historical overview of the evolution of the American approach to land warfare over the last 75 years reveals three broader trends important for framing a discussion of the lessons learned from the most recent conflicts.

- First, while it is almost hackneyed to say, land warfare is growing increasingly complex. Land warfare today relies on technologically sophisticated systems to collect, share, and process information and carry out precision strikes, capabilities that would have been unimaginable in the World War II period. As importantly, the conduct of land warfare is increasingly structurally complex as well. More and more frequently, the Army operates in conjunction with its sister services, other government agencies, and international and multinational partners. Successfully conducting future land campaigns will require navigating both the technological and structural complexities of this environment.
- Second, the Iraq and Afghanistan wars are part of a larger shift under way in land warfare away from conventional wars fought against state actors to unconventional conflicts fought against nonstate actors. World War II and the Korean War were almost exclusively conventional conflicts, at least from the perspective of the U.S. Army, and the U.S. Army did play a major role in the occupation and reconstruction of postwar Europe and Japan. However, the Vietnam War showed that much more emphasis had to be placed on irregular warfare. In both Iraq and Afghanistan,

the conventional phase was very short—a matter of a few weeks. The counterinsurgency period, however, lasted years. While Iraqi insurgents, the Afghan Taliban, and al Qaeda have far less military capability than a well-armed nation state, they present a different set of challenges to the U.S. military.[66]

- Finally, and perhaps most importantly, the Army traditionally has been more adept at learning tactical and operational lessons, rather than at learning (or recognizing) fundamental shifts in warfare. Again and again, to its credit, the Army embraced new technologies relatively quickly over the course of the years to address tactical and operational challenges and opportunities—like the helicopter, precision weapons, and more advanced intelligence and surveillance equipment. At the same time, the Army (and most of the joint force) has found it more difficult to make bigger-picture, strategic changes, particularly those that require capabilities other than those needed for conventional warfare.

Together, these three trends—the increasing JIIM nature of land warfare, the trend toward unconventional warfare, and the challenges the Army confronts at learning these big lessons about conflict—help give the remainder of this work its focus. Instead of focusing on the tactical or technological lessons learned from the last dozen years of war, the work focuses on the areas where the Army should seek to improve its institutional learning—on the strategic and policy levels of war. In essence, then, this work seeks to address Secretary Gates' twofold challenge of avoiding both "next-war-itis," as well the Army's tendency to focus strictly on high-end conflicts. In sum, it seeks to help the Army learn big.

---

[66] Thomas S. Szayna, Angela O'Mahony, Jennifer Kavanagh, Stephen Watts, Bryan A. Frederick, Tova C. Norlen, and Phoenix Voorhies, *Conflict Trends and Conflict Drivers: An Empirical Assessment of Historical Conflict Patterns and Future Conflict Projections*, unpublished RAND Corporation research, 2013, p. 30.

# CHAPTER THREE

# Lessons from 13 Years of War

The preceding chapter described the evolution of warfare as conducted by the joint force and particularly the land forces. This survey of the joint force experience would be incomplete and would be an insufficient basis for drawing lessons without the addition of two dimensions: the policy and strategy formulation and adaptation process and the interagency experience. The joint force does not operate without guidance from civilian policymakers, and without interagency, intergovernmental, and multinational partners. There has been no government-wide effort to synthesize lessons from the past 13 years at the policy level or with interagency input. The joint staff's *Decade of War* study relied on 46 studies conducted by the JCOA and a working group convened by the J7, but, as it acknowledged, that did not include the strategic, policy level, or interagency perspective.

This chapter builds on that work and represents an initial effort to begin to fill that gap. Between 2001 and 2014, the United States deployed upward of 250,000 troops to overthrow two governments, undertake protracted counterinsurgencies, conduct counterterrorism operations worldwide, and establish a global network of partner forces for counterterrorism operations. Examination of these years, one of the longest periods of conflict in U.S. history, should yield useful insights for future conflicts. A full analysis of lessons from the recent conflicts must await comprehensive case studies that describe the full narrative of events, explain their outcomes, and project their relevance for future conflict.

This chapter draws heavily on a workshop of scholars and policymakers convened by RAND to begin the work of assessing the recent period of war, and thereby advance the understanding of how the U.S. government may use all instruments of national power more effectively. There have been few comparable efforts by experienced policymakers and academic experts in recent years, making this workshop uniquely valuable for its insights. In order to identify critical lessons from the past 13 years and potential reforms of policy, practices, and/or organizations to address them, the workshop was designed in three modules of structured conversations on policy, strategy, and implementation.

## Findings

The workshop participants identified a wide range of problems with, and solutions to, the United States's defense and national security policymaking process. For purposes of clarity, we have grouped their insights into seven lessons. These lessons incorporate the comments and insights of the workshop participants, augmented by relevant external research, RAND analysis, and the findings of the other chapters in this report. These lessons do not, however, necessarily reflect the views of any workshop participant, whose comments were given off the record in a not-for-attribution context. Many of these issues predate 2001 and have been the subject of considerable debate and scholarship. In some cases the debate has been somewhat settled, and in other cases opposing camps continue to press their case. In some cases agreement was reached on the need for a remedy, but the remedy was inadequate or was not implemented consistently or successfully. In yet other cases, the centrality of the issue has not been widely acknowledged. This study identifies the contrasting positions on each issue, examines them in light of the recent record, and makes a case for their continued relevance and need for remedial action.

### 1. The Making of National Security Strategy Has Suffered from a Lack of Understanding and Application of Strategic Art

A major point of discussion among workshop participants was the degree to which the U.S. government struggled with crafting policy and strategy in the past 13 years.[1] Very often, the policy decisions and strategy on major issues of the wars did not produce good outcomes, as illustrated in the examples summarized below. A number of complex issues account for the difficulty. First, civilian policymakers and the U.S. military have different conceptions of how policy and strategy should be made. Second, policymakers have a tendency to eschew strategy and focus on tactical issues. Third, and perhaps most important, is a desire to pursue a technocratic approach to strategy that achieves tactical and operational successes without securing the ultimate objectives sought. Finally, policymakers and military leaders may not see strategy as essentially an adaptive art for coping with the uncertainties of war and the lack of perfect knowledge. A significant body of scholarship has identified these issues, and some effort has been made to increase and improve education in strategy, but a wider appreciation of the degree to which this deficit produces suboptimal national security outcomes may be lacking.

Regarding the civil-military relationship, one position is that making strategy should be a linear process, with civilian policymakers setting objectives and the U.S. military leadership crafting the strategy to achieve them. Another camp holds that in practice the line between policy and strategy is actually very blurry, and that civilians and the military need to engage in an iterative and dynamic dialogue to inform policy decisions and subsequent strategy. A third view holds that the approach to strategy will vary greatly according to presidential preference. Given that the U.S. presidential system grants substantial authority for making foreign policy to the chief executive, within the bounds of the Constitution and U.S. law, national security decisions and the manner of making them will be determined in some measure by the sitting U.S. President. The President is ultimately the decisionmaker,

---

[1] The military defines strategy as "a prudent idea or set of ideas for employing the instruments of national power in a synchronized and integrated fashion to achieve theater, national, and/or multinational objectives" (U.S. Department of Defense, *Dictionary of Military and Associated Terms*, Joint Publication 1-02, 2014, p. 244). Essential to the concept of strategy is the process of identifying priorities, articulating assumptions, making decisions, eliminating options, bringing goals in line with available resources, and choosing what not to do as much as what *to* do.

the nominator of his team, and the architect of his process for making decisions. Yet the experience of the past 13 years suggests that effective civilian and military interaction is critical to the framing of realistic policy objectives and effective strategy for their achievement. The development of policy objectives, policy options, and strategy requires (1) an accurate characterization of the conflict and the adversary; (2) an understanding of the possible ways of addressing the problem, with the attendant risks and assumptions; and (3) an estimate of means and time required to execute those possible ways. The military is an essential provider of those inputs, as are the intelligence community and others, to assist in the framing of objectives and the assessment of options. The dynamic dialogue requires a degree of trust and interaction in an iterative process.[2]

This view stands in contrast to the formal strategy-making process framed in U.S. military doctrine and taught in professional military education. In this view the military is given guidance from policymakers and then crafts strategy that aligns ends, ways, and means. Joint doctrine states that the President "establishes policy and national strategic objectives," after which the Secretary of Defense "translates these into strategic military objectives" and combatant commanders then identify "the military end state" through "theater strategic planning." Military doctrine describes the White House's national security strategy (NSS) as "a broad strategic context for employing military capabilities in concert with other instruments of national power. In the ends, ways, and means construct, the NSS provides the ends."

Numerous scholars have argued that the demands of the contemporary security environment require a broader understanding of strat-

---

[2]   Two recent examinations of difficulties in the formulation of U.S. strategy are Andrew F. Krepinevich, Jr., and Barry D. Watts, *Regaining Strategic Competence*, Washington, D.C.: Center for Strategic and Budgetary Assessments, 2009, and Francis G. Hoffman, "Enhancing America's Strategic Competency," in Alan Cromartie, ed., *Liberal Wars*, London: Routledge, forthcoming. See also Project on National Security Reform, *Forging a New Shield*, 2008; Clark A. Murdock et al., *Beyond Goldwater-Nichols: Defense Reform for a New Strategic Era*, Washington, D.C.: Center for Strategic and International Studies, Phase 1 Report, 2004, and Phase II Report, 2006; Steve Metz, *Strategic Landpower Task Force Research Report*, Strategic Studies Institute, October 3, 2013; and James Burk, *How 9/11 Changed Our Ways of War*, Stanford, Calif.: Stanford University Press, 2013.

egy[3] and, thus, a different relationship between civilians and the military in the strategy-making process. As Eliot Cohen has written, "both groups must expect a running conversation in which, although civilian opinion will not usually dictate, it must dominate; and that conversation will cover not only ends and policies, but ways and means."[4]

Sir Hew Strachan has also stated that this issue is fundamentally about getting the civil-military relationship right: "Even if America and Britain were clearer about strategy as a concept, they would still not be able to say definitively which governmental body makes strategy. This . . . question is in the first instance a matter of civil-military relations. The principal purpose of effective civil-military relations is national security: its output is strategy. Democracies tend to forget that. They have come to address civil-military relations as a means to an end, not as a way of making the state more efficient in its use of military power, but as an end in itself. Instead the principal objective, to which others become secondary, has been the subordination of the armed forces to civil control."[5]

The degree to which the issue of civil-military relations has dominated and created friction in the making of policy and strategy in the past 13 years is a major theme in the memoirs of former Secretary of Defense Robert Gates. He viewed a number of senior White House officials as inexperienced in national security matters and criticized the military for behavior that civilians perceived as boxing the president in.[6] He attempted at several junctures to bridge the gap and did so in brokering compromise positions on timelines and glide slopes for the withdrawal from Iraq and the transition in Afghanistan.

---

[3]  See, for example, Colin S. Gray, *The Strategy Bridge*, New York: Oxford University Press, 2010, and Hew Strachan, *The Direction of War: Contemporary Strategy in Historical Perspective*, New York: Cambridge University Press, 2013.

[4]  Eliot A. Cohen, *Supreme Command: Soldiers, Statesmen, and Leadership in Wartime*, New York: Free Press, 2002. See also Peter Feaver and Richard H. Kohn, eds., *Soldiers and Civilians: The Civil-Military Gap and American National Security*, Cambridge, Mass.: MIT Press, 2001.

[5]  Strachan, 2013, p. 76.

[6]  Robert M. Gates, *Duty: Memoirs of a Secretary at War*. New York: Alfred A. Knopf, 2014, pp. 301, 338–339, 352, 365–369, 381–383.

The second major issue is the locus of strategy-making, with several participants and a burgeoning literature noting that the White House and senior civilian policymakers have eschewed the making of strategy while becoming embroiled in operational and tactical details. Gates documents the high degree of focus in the White House on such tactical details as dates and troop levels, which often consumed the attention of the most senior officials rather than the critical war-ending issues of political outcomes, negotiations, and elections, where civilian insights are critical. Numerous workshop participants with exposure to the interagency and White House deliberations concurred with this characterization.

The third reason for the strategy failings may not be trivial flaws of process or friction in the bureaucracy but a narrow view of strategy and a narrow view of war. The fundamental problem could be, as Colin Gray has argued, rooted in what he calls "the American way of war" (referencing and seeking to update, no doubt, Russell Weigley's superb history of American military strategy and policy[7]), which was laid bare and exacerbated by the challenges posed by nonstate adversaries using asymmetric tactics to capitalize on the attendant U.S. blind spots and vulnerabilities.[8] In essence, he charges that the United States has an overly narrow conception of war that does not account for its full dimensions and thus hobbles its making of both policy and strategy: "The United States has a persisting strategy deficit. Americans are very competent at fighting, but they are much less successful in fighting in such a way that they secure the strategic and, hence, political rewards they seek. The United States continues to have difficulty regarding war and politics as a unity, with war needing to be permeated by political considerations. American public, strategic, and military culture is not friendly to the means and methods necessary for the waging of warfare against irregular enemies. The traditional American way of war

---

[7]   Russell F. Weigley, *The American Way of War: A History of United States Military Strategy and Policy*, Bloomington, Ind.: Indiana University Press, 1977.

[8]   Colin S. Gray, "Irregular Enemies and the Essence of Strategy: Can the American Way of War Adapt?" Carlisle, Pa.: Strategic Studies Institute, 2006.

was developed to defeat regular enemies."[9] So as not to argue against a straw man, Gray identifies and describes what he considers to be the chief characteristics of the American way of war—namely, that it is apolitical, astrategic, ahistorical, optimistic, culturally challenged, technology dependent, focused on firepower, large scale, offensive, profoundly regular, impatient, logistically excellent, and highly sensitive to casualties.[10]

The final element that may be missing from the U.S. conception of strategy is a recognition that it must be iterative and adaptive due to the very real possibility that assumptions may prove to be wrong and the fact that war, as with all human affairs, is inherently uncertain. Therefore, there must be an expectation and willingness to reexamine and acknowledge flawed assumptions and undesirable outcomes, which, of course, is unpalatable, is politically risky, and can provoke an "ostrich syndrome" until disaster is on the doorstep. In the past 13 years, review and adaptation of policy and strategy was painful, belated, and often incomplete. Many participants noted that once policy was set, it was very difficult to revisit positions taken. It may be easier to accept the need to adjust course if adaptation is rooted in the concept of strategy. As Strachan writes, "Strategy occupies the space between a desired outcome, presumably shaped by the national interest, and contingency, and it directs the outcome of a battle or of another major event to fit with the objectives of policy as best it can. It also recognizes that strategy may itself have to bend in response to events. Essential here is the need for flexibility and adaptability. Thus strategy must be viewed as the activity that that "offers options[,] not a straightjacket."[11]

While many of these insights can be traced to Clausewitz, that timeless quality does not make them invalid or unfruitful to apply to recent experience. Indeed, the enormity of the U.S. endeavor over the past 13 years and the prospect of a future riddled with continued threats and more constrained resources make it imperative to assess the profoundest roots of the shortcomings. More than ever, the United

---

[9]  Gray, 2006, pp. vi–vii.

[10]  Gray, 2006, pp. 30–48.

[11]  Strachan, 2013, pp. 251, 266–267.

States appears in need of some useful guideposts for crafting objectives that are realistic, ways that stand a reasonable chance of achieving those ends, with a proportionate outlay of resources over time. The following four examples of U.S. policy and strategy in the past 13 years illustrate the difficulties described in general terms above.

- The decision to go to war in Iraq was notable among the major policy decisions of the past 13 years for relying on very little process or structure, as major figures in the Bush administration persuaded the president to invade Iraq without any extended deliberation.[12] The intelligence failures of this period have been amply documented; the feared existence of weapons of mass destruction in Saddam Hussein's Iraq did not pan out. Intelligence is not perfect and can never provide total certainty, but a more important failure in the decisionmaking process was the exclusion of cautionary or contrary views from senior State Department officials and a State Department assessment led by Tom Warrick warning of the problems that could ensue in Iraq as a consequence of a military intervention to remove Hussein from power. The assumptions that proved to be wrong included the expectation that the intervention would be quickly concluded, that troops would not be needed to conduct stabilization and reconstruction, that political factions in Iraq would readily cooperate to form a new government that would be reasonable functional, and indeed that the basic infrastructure of Iraq would function.[13] Relying on the experience of the Balkans, Army chief of staff Gen. Eric Shinseki had testified to Congress on

---

[12]   Bob Woodward, *Plan of Attack*, New York: Simon & Schuster, 2004. Also, regarding the decision to go to war in Afghanistan, in contrast with many observers who believe that the exigency of responding to the 9/11 attack excused the lack of a strategy, Strachan writes: "[Defense secretary] Rumsfeld's response was to bypass the problem of strategy, not confront it. In 2001–02, he, and with him the President and the vice-president, marginalized the Chairman of the Joint Chiefs of Staff. . . . The Americans were saved from the consequences of their own temerity because of the sudden and wholly unexpected collapse of Kabul, thanks to the military contribution of the Northern Alliance" (Strachan, 2013, p. 68).

[13]   Nora Bensahel, Olga Oliker, Keith Crane, Rick Brennan, Jr, Heather S. Gregg, Thomas Sullivan, and Andrew Rathmell, *After Saddam: Prewar Planning and the Occupation of Iraq*, Santa Monica, Calif.: RAND Corporation, MG-642-A, 2008, p. 233.

February 25, 2003, that several hundred thousand soldiers would be needed for post-combat stability operations, but that view was firmly rejected by the U.S. Department of Defense (DoD) civilian leadership. In the event, the military was left unprepared not only for that task but for the job of rebuilding a new army and police force, which was disbanded in an effort to "de-Baathify" both the military and the government. These steps were interpreted by Iraqi Sunnis as an attack on them, and together with Shia sectarianism, the fires of insurgency were ignited before the year was out.

- The next major decision was the decision to send a "surge" of U.S. forces into Iraq in 2007 to turn around what had become a full-blown sectarian civil war between Shia and Sunni Iraqis, with a virulent strain of al Qaeda finding sanctuary and recruits within a disaffected Sunni population. The two notable features of this decision were that it took over three years of spiraling violence for policymakers to come to grips with the situation and act, and that ultimately its decision was made by going outside the formal civil-military process for advice and relying on advice contrary to that given by the senior military leadership at the Pentagon and U.S. Central Command. Regarding the first point, LTG H.R. McMaster, a participant in both wars and a scholar whose doctoral dissertation examined the civil-military frictions and role of the joint chiefs in the Vietnam war, located the root of this slow adaptation of strategy in the still-potent U.S. vision of relying on technology to wins its wars through the network-centric "revolution in military affairs." In his view, "the disconnect between the true nature of these conflicts and pre-war visions of future war helps explain the lack of planning for the aftermath of both invasions as well as why it took so long to adapt to the shifting character of the conflicts after initial military operations quickly removed the Taliban and Ba'athist regimes from power."[14] Regarding the second point, political scientist Peter Feaver describes the process by which the Bush White House, in which he served, relied extensively on outside advisers and ultimately rejected senior military leaders' rec-

---

[14]  H.R. McMaster, "On War: Lessons to Be Learned," *Survival*, Vol. 50, No. 1, 2008, p. 25.

ommendations. The President decided instead to send a surge of troops to Iraq, which helped induce Sunnis to turn their arms against al Qaeda in Iraq and resulted in a reduction of violence by the summer of 2008.[15] The President can certainly rely on outside advice, but the example reveals a civil-military dialogue that did not result in options that the President saw as meeting his objectives.

- The third major decision of the past 13 years was the decision to follow suit in Afghanistan and send a surge of U.S. troops to address spiraling violence there. In contrast to the limited deliberations in the decision to invade Iraq and the influential outside advice that carried the day in the Iraq surge decision, this decision was preceded by several lengthy policy reviews carried out by the Bush and then Obama White Houses, an assessment by U.S. Central Command (CENTCOM), and then an assessment undertaken by General Stanley McChrystal that purported to gauge the number of troops and other resources ("means") needed but in fact reopened the entire question of the approach to the war. In 2006, 2008, and 2009, the National Security Council (NSC) hosted a series of interagency reviews of policy and strategy towards Afghanistan and Pakistan.

While less widely heralded than the review of Iraq policy that led to the surge, the series of Afghanistan-Pakistan reviews did take a holistic, regional approach and attempted to formulate a whole-of-government to the conflict. The 2006 review resulted in a dramatic increase in U.S. aid, especially security assistance, to Afghanistan. The subsequent appointment of a Deputy National Security Advisor for Iraq and Afghanistan helped bring more attention, coordination, and integration to the war efforts. In 2008 and 2009, the reviews were accompanied by parallel reviews by the Joint Staff and CENTCOM and, later, by General Stanley McChrystal when he assumed command of U.S. forces in theater. The reviews typically involved two dozen participants from across

---

[15]  Peter Feaver, "The Right to Be Right: Civil-Military Relations and the Iraq Surge Decision," *International Security*, Vol. 35, No. 4, Spring 2011. See also Woodward, 2004.

the agencies and departments meeting for three to four hours several times per week for three to four weeks. In addition to the regular participants, the review groups were briefed by outside experts from the intelligence community, allied governments, and even Afghan and Pakistani officials.[16] Officers from the Joint Staff and civilians from the Office of the Secretary of Defense did participate, but the military also conducted its own parallel reviews, leading to some disconnect. After the 2008 U.S. presidential election campaign, the incoming Obama administration initiated its own review.

What was the upshot of the reviews? As Gates recounts in his memoirs, he and his under secretary of defense for policy, Michele Flournoy, considered the 2009 review led by Bruce Riedel to have described "what but not how," thus not meeting the principal requirement of a strategy. The lengthy reviews and the eventual decision to send 30,000 more troops to Afghanistan for two years obscured the fact that consensus was never reached on the approach to the war: General McChrystal ascribed to a fully resourced counterinsurgency effort and the Obama White House sought a less ambitious effort focused on keeping al Qaeda from expanding its sanctuary in South Asia.

- The fourth major decision, or policy, enacted in the past 13 years was the counterterrorism policy adopted to fight the al Qaeda terrorist organization that attacked the United States on September 11, 2001, and affiliates that have sprung up in the ensuing years. The congressional authorization for the use of military force has been interpreted as permitting the U.S. forces to strike at any al Qaeda affiliate deemed to represent a dire or imminent threat to U.S. persons or the homeland. It has led to a continuous campaign of raids and drone strikes in several countries around the globe beyond the war theaters of Iraq and Afghanistan. Some scholars have defended the efficacy of this approach as preventing attacks on the homeland and U.S. citizens and avoiding the

---

[16]  Bob Woodward, *Obama's Wars*, New York: Simon & Schuster, 2011, pp. 43, 246.

expense or pitfalls of extended or large-scale operations.[17] This approach was born of a desire by the civilian leadership to find a cost-effective approach to degrading and disrupting if not destroying al Qaeda, concerns about counterproductive effects of large-scale military endeavors, and skepticism about the U.S. ability to affect underlying conditions or causes through more comprehensive approaches. Others have argued that this approach amounts to applying tactics in lieu of a strategy. As Audrey Cronin has written, this approach does not answer the question of how the war against al Qaeda will end; in other words, it does not articulate the policy objectives beyond "keeping pressure on the network" in a bid to prevent further attacks on the homeland.[18] If this is the strategy, then it is a prescription for endless war, or at least endless raids when intelligence suggests a threat is imminent. Because the strategy does not articulate how the al Qaeda threat will be removed or permanently resolved, there will always be in theory another possible attack to thwart. (This critique mirrors the one that has been made of the counterinsurgency approach applied in Iraq and Afghanistan, which will be discussed in Lessons 3 and 5. Counterinsurgency methods are indeed tactics that, if not harnessed to a strategy, are not capable of producing a lasting outcome.)

Counterterrorism has been defined in vastly different ways over the past 13 years. Military doctrine defines it as "actions taken directly against terrorist networks and indirectly to influence and render global and regional environments inhospitable to terrorist networks."[19] But

---

[17]   Daniel Byman, "Why Drones Work: The Case for Washington's Weapon of Choice," *Foreign Affairs*, Vol. 92, No. 4, 2013, p. 32.

[18]   Audrey Kurth Cronin, "The 'War on Terrorism': What Does It Mean to Win?" *Journal of Strategic Studies*, Vol. 37, No. 2, 2014, pp. 174–197. It should be noted that under the Bush administration a more comprehensive approach to the "global war on terrorism" was framed, to include an "indirect approach" that strengthened allies and aimed to deradicalize potential recruits, but this approach was not fully developed as a strategy or robustly resourced. Lesson 7 will address the case of the Philippines.

[19]   U.S. Department of Defense, 2014, p. 60.

in application it has often focused on the direct, enemy-centric part of that definition, which constitutes reliance on an even narrower set of tactics than counterinsurgency methods. McMaster and others have criticized the U.S. reliance on a minimalist-footprint counterterrorism approach. In his view, "This approach elevated one important capability in counterinsurgency to the level of strategy. It did not adequately address fundamental causes of violence, critical sources of enemy strength, the enemy strategy, likely enemy reactions, or the effect of the actions on the population."[20] The minimalism also extends to the definition of U.S. interests. The core objectives of keeping the U.S. homeland and U.S. citizens safe are by themselves a narrow framing of U.S. national security interests that does not adequately account for the fact that instability and sanctuaries in key areas of the world are the breeding grounds for those who would launch such attacks.

Interestingly, the "counterterrorism only" approach, as it has been dubbed, was embraced more by civilians than the U.S. military, including SOF that would be one of the primary entities prosecuting such an approach. In order to dissuade senior White House officials, Gates commissioned a paper from McChrystal, who had led the counterterrorism special operations task force, to explain in some technical detail why this "CT only" approach would not work in Afghanistan. He wrote that "without close-in access, fix and find methods become nearly impossible. Predator strikes are effective where they complement, not replace, the capabilities of the state security apparatus, but they are not scalable in the absence of underlying infrastructure, intelligence, and physical presence."[21] Targeting terrorists has been the main thrust of U.S. counterterrorism policy, but it has also included efforts to prevent and counter radicalization and improve homeland defense, border control, and law enforcement in the U.S. and abroad, as well as significant intelligence reform to encourage greater cooperation among the many agencies of the intelligence community.

---

[20]  H.R. McMaster, "Decentralization vs Centralization," in Thomas Donnelly and Frederick W. Kagan, *Lessons for a Long War: How America Can Win on New Battlefields*, Washington, D.C.: AEI, 2010, pp. 64–92.

[21]  Gates, 2014, p. 364.

What have been the results? While the counterterrorism approach as applied decapitated much of the original al Qaeda leadership, the leadership subsequently regenerated and the movement metastasized. As the director of national intelligence testified in 2014 in presenting the annual worldwide threat assessment, "The threat of complex, large-scale attacks from core al-Qa'ida against the U.S. Homeland is significantly degraded. . . . However, diffusion has led to the emergence of new power centers and an increase in threats by networks of like-minded extremists with allegiances to multiple groups."[22] No attacks against the U.S. homeland have been successful, but particularly since the declaration of an Islamic State caliphate in substantial parts of Iraq and Syria in mid-2014, U.S. officials acknowledge the end of the counterterrorism fight is not in sight.

## 2. An Integrated Civilian-Military Process Is a Necessary, But Not Sufficient, Condition of Effective National Security Policy and Strategy

The need for an effective, integrated process for making policy and strategy has been widely recognized, and various directives have sought to establish procedures, but current practices appear less rather than more functional. Moreover, a lack of trust and skepticism about efficacious ways to achieve objectives at reasonable cost have complicated the civil-military dialogue. One workshop participant noted that in his experience of administrations since the Vietnam War, three factors affected the conduct of national security strategy: the President's own leadership, the experience of his team, and the interaction of the principal members, which in turn affected whether the established process worked and produced sound policy and strategy. The George H. W. Bush national security cabinet included significant experience in previous Republican administrations, relative harmony among the principals, and a National Security Adviser who played the role of arbiter in a process that was considered inclusive and orderly.

---

[22] James R. Clapper, "Statement for the Record: Worldwide Threat Assessment of the U.S. Intelligence Community," Senate Select Committee on Intelligence, January 29, 2014, p. 7.

As noted above, some scholarship views the presidential decision-making process as overwhelmingly a function of presidential personality—which implies that there is little point to reforming the process.[23] In contrast to this view, while stipulating that presidential preference is a critical factor in determining what decisions are made and how, most workshop participants and other scholarship suggest that structure and process can make an enormous difference.[24] Many factors have affected the working of the national security policy and strategy-making process since the first Bush administration. A key factor identified is the fact that the NSC staff has ballooned to some 500 members, which makes it a rival center of policy analysis and fosters a tendency for this body to take positions rather than organize and arbitrate a considered review of the options generated by the principals. In run-up to the war in Iraq, for example, a RAND study found that "The NSC seems not to have mediated the persistent disagreement between the Defense Department and the State Department that existed throughout the planning process."[25] By taking a position the NSC can also become wedded to a course when it should reevaluate the options. Finally, as the NSC has become increasingly occupied with operational matters, it has less ability to maintain a strategic perspective.

---

[23]  Richard E. Neustadt, *Presidential Power*, New York: New American Library, 1960. I. M. Destler, "National Security Advice to US Presidents: Some Lessons from Thirty Years," *World Politics*, Vol. 29, No. 2, 1977, pp. 143–176. George C. Edwards, III, and Stephen J. Wayne, *Presidential Leadership: Politics and Policy Making*, New York: Worth Publishers, 1999. For the opposite view, see Amy Zegart, *Flawed by Design: The Evolution of the CIA, JCS, and NSC*, Stanford, Calif.: Stanford University Press, 1999.

[24]  H.R. McMaster, "On War: Lessons to Be Learned," *Survival*, Vol. 50, No. 1, 2008, pp. 19–30; Eliot A. Cohen, "The Historical Mind and Military Strategy," *Orbis*, Vol. 49, No. 4, 2005, pp. 575–588; Janine Davidson, "The Contemporary Presidency: Civil-Military Friction and Presidential Decision Making: Explaining the Broken Dialogue," *Presidential Studies Quarterly*, Vol. 43, No. 1, 2013, pp. 129–145; H.R. McMaster, "Effective Civilian-Military Planning," in Michael Miklaucic, ed., *Commanding Heights: Strategic Lessons from Complex Operations*, NDU Press, 2010; Thomas S. Szayna, Kevin F. McCarthy, Jerry M. Sollinger, Linda J. Demaine, Jefferson P. Marquis, and Brett Steele, *The Civil-Military Gap in the United States: Does It Exist, Why, and Does It Matter?* Santa Monica, Calif.: RAND Corporation, MG-379-A, 2007.

[25]  Bensahel et al., 2008, p. 238.

The Interagency Policy Committee (IPC), Deputies and Principals meetings process was intended to create the forum for interagency deliberation and vetting of policy options for eventual presidential decisions. Participants observed that even senior officials have become consumed by details and crisis management, an observation also made repeatedly by Gates, who served eight administrations.[26] Principals' growing number of direct reports also increases their management burden and reduces the time for strategic thought. Workshop participants with experience in the policy process relate the endless cycle of preparing for meetings, briefing their principals for meetings, and sitting in meetings. The time for in-depth thought in the U.S. bureaucracy is shockingly limited.[27]

Another phenomenon noted is compartmentalization, which especially afflicted counterterrorism policy. A very small group of officials made decisions on targeting al Qaeda, but often in the absence of broader policy discussions regarding the country or region in question. The counterterrorism strategy was treated as apart from the broader national security strategy and the policies toward the countries or regions in question. The focus on operations increased the CIA's focus on tactical rather than national intelligence. Yet its director was called upon to determine what constituted success in the war on terrorism, which is a policy and strategy issue.[28]

Many of these systemic shortcomings were documented in the voluminous Project on National Security Reform, but the many reforms proposed were not adopted.[29] Workshop participants agreed that process problems were fundamental and if addressed could enhance the making of policy and strategy. Three requirements were identified.

---

[26]  Gates, 2014, pp. 352, 371.

[27]  This observation was also documented in an earlier study, Clark A. Murdock et al., *Beyond Goldwater-Nichols: Defense Reform for a New Strategic Era*, Phase 1 Report, Washington, D.C.: Center for Strategic and International Studies, March 2004, pp. 61–62.

[28]  "Panetta: 'My Mission Has Always Been to Keep the Country Safe,'" National Public Radio, February 3, 2013.

[29]  Center for the Study of the Presidency & Congress, Project on National Security Reform, "Forging a New Shield," November 2008.

First, the process must be interactive and dynamic, not linear. Second, the process must integrate civilians and the military. Finally, an adequate process must revisit assumptions, estimate reactions and unintended effects, and enforce a review and revision cycle. Participants noted that hearing a diversity of views was vital, including a more articulated expression of the intelligence analysis that included divergent views and acknowledgment of gaps. Visiting the country in conflict, meeting with the locals, and surveying the battlefield provide senior leaders an invaluable window into the firsthand realities that will help them craft, assess, and adjust their thinking and subsequent strategy.

The making of good strategy and policy is impeded by the gulf between the civilian and military view of how the process should work. The difference in views regarding strategy as discussed in the previous lesson is mirrored in a disjunction between the military approach to the process and the needs of civilians charged with making policy decisions. Military doctrine teaches that national security policy objectives are formulated and then translated into the National Security Strategy, the National Military Strategy, the Quadrennial Defense Review, and regional campaign plans, and thence down to specific operational and tactical plans. But determining objectives requires input from the military that will inform the President's decision. As one workshop participant described it, "The President will say, 'Here is the problem. I want some thoughts. Please give me some options.' The military wants to know what the end state is. The President needs to weigh other objectives, resources needed, the politics of his decision, etc. The military feels that they aren't getting sufficient guidance, and the President feels he isn't getting clear options."

As Janine Davidson has observed, based on three years as the deputy assistant secretary of defense for plans, the military's planning process is at odds with the needs of the civilians and the requirements of good policy: "Military planners want detailed guidance regarding end states and objectives that civilians often cannot provide up front. . . . Civilian presidents and defense secretaries might be surprised to learn that it is considered their responsibility to determine ends and means without being first offered a menu of feasible options."[30] After 2001, an adaptive planning process was implemented that instituted regular civilian review of military plans, but it requires an open dialogue to function. It also does not provide a sufficiently flexible and rapid model for strategy making in crisis situations.

Moreover, there is a deep source of the tension that process alone may not be able to resolve. The military leader is legally obligated to provide his best professional advice (i.e., accurate to the best of his knowledge) and an executable plan, but that imperative can be wielded in a way that restricts the presidential prerogative to set policy.[31] This friction was on display in the debate over Afghanistan, as noted above. The President felt "boxed in" by the military proposal to send 40,000 more troops to Afghanistan as the only real option, while the military felt that their advice regarding the means needed to achieve the objectives was being disregarded. The ingrained tension stems from the fact that the military will generally prefer to have more resources and more time to maximize the chance of success, while the President may be inclined to seek a more minimalist approach, given the range of other considerations he/she must weigh. The problem is not that the President may disagree with his military advisers and reject their advice—that is the President's prerogative, not a flaw in the process—but that the President and his military advisers believe different things about their respective roles and responsibilities in the decisionmaking process. The sense that civilians are not competent to make strategy can lead the military to believe that only it has the knowledge to understand the correct course, which is the essence of a civil-military crisis. Davidson argues that a version of this crisis of confidence is embedded in the Powell Doctrine, which purports to stipulate rules for when force

---

[30]   Janine Davidson, "The Contemporary Presidency: Civil-Military Friction and Presidential Decision Making: Explaining the Broken Dialogue," *Presidential Studies Quarterly*, Vol. 43, No. 1, 2013, p. 141.

[31]   Peter Feaver, "Crisis as Shirking: An Agency Theory Explanation of the Souring of American Civil-Military Relations," *Armed Forces & Society*, Vol. 24, No. 3, 1998, p. 415.

should be used.[32] This "broken dialogue" cannot be mended by process alone, but process can help clarify roles in strategy formulation.

The second issue identified is the need for an integrated process. Once the presidential decision has been reached, the civilian role in strategy does not end, but there is no mechanism for integration below the level of the cabinet. Each department is empowered to carry out its own portfolio, and there is no forcing function built into the government for either planning or execution in concert. This is not a new issue; the need for an integrated civil-military process to craft the strategy and the plans that flow from it (as well as to work together in implementation as will be addressed in Lesson 7) has been widely recognized. Solutions have been attempted at various times. One such mechanism was promulgated in Presidential Decision Directive 56, implemented in the Clinton administration after the failures of Somalia in the early 1990s.[33] PDD-56 required departments and agencies to review their legislative and budgetary authorities to ensure they had sufficient capabilities and resources to support complex operations. It envisioned the creation of an executive committee by the deputies to manage complex operations and oversee the development of an interagency political-military implementation plan. It called for an interagency "rehearsal" of the political-military plan by the departments and agencies to test its viability before a major operation is undertaken. PDD-56 was not implemented by successive administrations, and the successor document in the Bush administration (National Security Presidential Directive 44) for stabilization and reconstruction did not overcome the challenge that no department can dictate to another. Mandated interagency coordination for all complex political-military operations with a presidentially designated lead is the only way to ensure that the necessary integrated civilian-military planning does occur. The establishment of such a formal requirement would need to be accompanied by other steps to introduce and foster a culture of planning within the civilian government.

---

[32] Davidson, 2013, p. 143.

[33] The Clinton administration issued Presidential Decision Directive 56 (PPD), "Managing Complex Contingency Operations," in May 1997.

The third issue identified is a process that is sufficiently sophisticated to take account of the many variables involved in making strategy. The U.S. Army has adopted the tenets of design to improve upon the linear military decision making process. A much-simplified version of systemic operational design as originated by the Israeli military, this approach dictates an initial assessment to accurately diagnose, or "frame" the problem, and an iterative process of establishing an approach and then adjusting the approach based on the results achieved and the changing environment. One example of this approach was performed by a multinational expert group led by McMaster and U.S. diplomat David Pearce to assess the cause of the failing war in Iraq and suggest a new approach. In 2007, General David Petraeus assembled a Joint Strategic Assessment Team (JSAT) to evaluate the war in Iraq and recommend a way forward. Chaired by H.R. McMaster (then a U.S. Army colonel), the JSAT was an interagency, civil-military, multinational review team of some two dozen personnel that included representatives of the U.S. State Department, think tanks, and allied governments.

The JSAT sought to understand and diagnose the current state of the war in Iraq and then recommend an appropriate strategy. The team found that the war in Iraq had become a sectarian civil war and that among other things political reconciliation would be required to provide a stable government. If the Iraqi government members were not able to see this as in their interests because they favored their particular sect or a winner take all approach, the approach condoned efforts to marginalize sectarian leaders and neutralize their influence. The JSAT recommended a CORDS-like structure with a single leader to achieve a tight integration of this political-military strategy, which was not adopted. But closer coordination was achieved between civilians and the military in Iraq, starting with Petraeus and Ambassador Crocker. However, the approach was not adopted or implemented at the White House level as the overall strategy, and it was not continued by the succeeding administration.[34]

---

[34]  Robinson, 2008, pp. 98ff.

The JSAT ideally would have been conducted at the outset of the war rather than in 2007 when the situation was dire. It would also be beneficial to apply this approach at the policy level. Frank Hoffman has suggested that "[t]he application of Design at the strategic level may afford the interagency community a more comprehensive methodology to resolve the complex interactive variables that constitute the art of strategy at the national level."[35] The U.S. military devotes an enormous amount of time and effort to gaming, simulations, and exercises that involved nonmilitary participants, sometimes including very senior officials. But that is not the same as a policy-level exercise organized by and for the civilians charged with making policy decisions and setting the parameters for strategy.

Some efforts have been made to institute combined planning practices, even though no directive has made it obligatory. The Senior Representative for Afghanistan and Pakistan (SRAP) staged several "rehearsal of concept" drills with a wide array of participants, to determine the best modes for executing the civilian-military strategy. Workshop participants noted that the Obama administration conducted tabletop exercises at the senior level on Iran and other issues, in a partial application of this approach, to assess potential reactions and effects to policy options.

Over the past two years another collaborative assessment, gaming and planning effort was undertaken by the U.S. Agency for International Development (USAID) and two combatant commands. In late 2012 U.S. Special Operations Command (SOCOM) and USAID started to collaborate on a Joint Sahel Project to develop a common understanding of conflict drivers in the Sahel region of Africa. Concerned with the rise of terrorist groups sympathetic to al Qaeda in the region—including al Shabaab in Somalia, Boko Haram in Nigeria, al Qaeda in the Islamic Maghreb in Algeria, and Ansar Dine in Mali—SOCOM and USAID understood that a solely military response would be as inadequate as a solely development response. A broader response

---

[35] Francis G. Hoffman, "Enhancing America's Strategic Competency," in Alan Cromartie, ed., *Liberal Wars*, London: Routledge, forthcoming.

would be required to foster stability in the region and long-term U.S. interests.

This project was undertaken against the backdrop of previous civil-military frictions on this topic, culminating in the coup in Mali and eventual intervention with the French in the lead. The experience laid bare the lack of a shared understanding within the U.S. government of the region, the threat and U.S. interests, as well as what the focus of U.S. efforts should be. That bedrock disjunction meant that there was no integrated plan, no consistent resourcing, and no metrics or attempts to judge the success or failure of the activities being undertaken.

To remedy these profound lapses, over the course of 18 months, the SOCOM-USAID collaboration widened to include partners from across the U.S. government. USAID taught its Interagency Conflict Assessment Framework to U.S. Africa Command (AFRICOM) and SOCOM, enabling them to complete an initial Sahel Desk Study. AFRICOM hosted the first of two war games, adapting its approach to fully integrate social, political, economic, and cultural variables into the war game framework. In May 2014 USAID hosted a first-of-its-kind Development Game. The extended collaboration, initiated from the staff level without senior policymaker guidance, resulted in the inputs for a new interagency, civil-military plan for U.S. engagement in North Africa.[36]

## 3. Military Campaigns Must Be Based on a Political Strategy, Because Military Operations Take Place in the Political Environment of the State in Which the Intervention Takes Place

Military campaigns take place in the social, cultural, and political contexts of the states in which they are fought, and any successful operation will be cognizant of those contexts and have a plan for how to operate within, exploit, influence, and achieve victory in them.[37] The

---

[36] USAID-USSOCOM, *Joint Sahel Project: Development Game After Action Report (AAR)*, May 27, 2014.

[37] Rupert Smith, *The Utility of Force: The Art of War in the Modern World*, New York: Random House, 2008.

lack of a political strategy, the failure to recognize its centrality, and its inadequate integration and sustained application may be the most important insights to arise from this inquiry.

At one level, the centrality of politics to war is widely recognized. The political nature of war has been articulated by Clausewitz, Sun Tzu, and others and is part of the long tradition of diplomacy, particularly the school of realpolitik diplomacy.[38] While military and other students of strategy and political theory may know these texts by heart, however, U.S. military doctrine and U.S. strategy documents routinely fail to articulate what a political strategy entails and how to systematically produce one. The need to fashion a political strategy has not been widely acknowledged in policy circles or military doctrine or education. It has not been embraced or systematically incorporated into the making of policy and strategy over the past 13 years. Desired political endstates such as the creation of a democratic system in Iraq or Afghanistan may be articulated, but a profound understanding of the specific political conditions and a detailed view of how they might be shaped is not part of strategy-making as currently practiced. Thus, for example, in Iraq policymakers relied on their own predilections, expatriates' views, and outside academics' ideas about democratization, Shia politics, and possible parallels between de-Nazification and de-Baathification, but did not subject the critical topic of the post-Saddam political order to sustained discussion or rigorous examination. Relevant expertise within the government was ignored. In hindsight it is obvious that the United States did not adequately consider the ramifications of unseating a dictatorship based on a Sunni minority in a majority Shia country with a long-running separatist Kurdish movement. A political strategy was essential if a stable new political order was to take hold after the regime change. Instead, the actions and inactions of the U.S. forces and political leadership led to a full-blown sectarian civil conflict. In Afghanistan, a centralized political system was created for a country essentially composed of regional power centers, and the

---

[38] Otto Von Bismarck, *Bismarck: The Man and the Statesman*, Volume 1, New York: Cosimo, Inc., 2005; Hans J. Morgenthau and Kenneth W. Thompson, *Politics Among Nations: The Struggle for Power and Peace*, New York: Knopf, 1978.

United States paid too little attention to tribal, ethnic, and regional factions and their impact on the resurgent Taliban.[39] The United States did not have accurate appreciation of the country's political dynamics and failed to forge an effective strategy with its putative ally, the government of Hamid Karzai.

The cause of this deficit, as noted in Lesson 1, appears to rest in the limited conception of strategy that both policymakers and strategists have embraced. Both the civilian and military U.S. leaderships tended to focus on combat operations and counterterrorism, drawing their objectives narrowly around defeating enemy forces and preventing al Qaeda–affiliated terrorists to lay claim to certain territory. U.S. forces did indeed successfully target and eliminate many insurgents and terrorists. But these kinetic operations in themselves were insufficient to enable the host governments to control its territory and borders, particularly when the actions of those governments were fueling the generation of terrorists, insurgents, and sympathizers.

Achieving the overall goal of the wars required a political strategy integrated with the military strategy. As H.R. McMaster argued:

> Because an insurgency is fundamentally a political problem, the foundation for detailed counterinsurgency planning must be a political strategy that drives all other initiatives, actions, and programs. The general objective of the political strategy is to remove or reduce significantly the political basis for violence. The strategy must be consistent with the nature of the conflict, and is likely to address fears, grievances, and interests that motivate organizations within communities to provide active or tacit support for insurgents. Ultimately, the political strategy must endeavor to convince leaders of reconcilable armed groups that they can best protect and advance their interests through political participation, rather than violence.[40]

---

[39] Thomas Barfield, *Afghanistan: A Cultural and Political History*, Princeton, N.J.: Princeton University Press, 2012; and Rajiv Chandrasekaran, *Little America: The War Within the War for Afghanistan*, New York: Vintage Books, 2012.

[40] H.R. McMaster, "Effective Civilian-Military Planning," in Michael Miklaucic, ed., *Commanding Heights: Strategic Lessons from Complex Operations*, Washington, D.C.: NDU Press, 2010, pp. 98–99.

McMaster has also formulated this thesis with regard not only to counterinsurgency but to war more generally: "If the indigenous government and its security forces act to exacerbate rather than ameliorate the causes of violence, the political strategy must address how best to demonstrate that an alternative approach is necessary to avert defeat and achieve an outcome consistent with the indigenous government's interests. If institutions or functions of the supported state are captured by malign or corrupt organizations that pursue agendas inconsistent with the political strategy, it may become necessary to employ a range of cooperative, persuasive, and coercive means to change that behavior and restore a cooperative relationship."[41]

A political strategy can employ a wide variety of ways to accomplish this. A political arrangement that serves the U.S. objectives must be a central feature of the strategy. Taking into account the centrality of the political dimension does not force the strategy into a prescribed course of action, but it does require that it be fully taken into account in the approach crafted. There are a variety of tools to address the political dimension. Any political strategy must recognize the agency and interests of the other parties. For example, power sharing and sharing resources (i.e., patronage) are time-honored means of achieving political stability. Channeling competition among groups into the political arena instead of the armed arena constitutes a successful outcome that can occur through negotiations, elections, or outright defeat and capitulation. In Iraq, the key focus of a political strategy would have been to strike a lasting agreement among Shia, Sunni, and Kurds (with Iranian acceptance or at least a plan to minimize its interference). In Afghanistan the focus of a political strategy would balance Pashtun, Tajik, and other major groups' concerns while reaching out to Pashtun sympathizers of the Taliban.

Efforts were made to pursue these political objectives at various times in the wars, and the critical necessity of achieving success in this regard was recognized by Ambassador Zalmay Khalilzad (in both

---

[41] H.R. McMaster, "Decentralization vs Centralization," in Thomas Donnelly and Frederick W. Kagan, *Lessons for a Long War: How America Can Win on New Battlefields*, Washington, D.C.: AEI, 2010, p. 84.

Afghanistan and Iraq), the Petraeus/Crocker duo, and the coalition command in Kandahar in 2010–2011 and in regional overtures by SRAP, but the absence of a clear consensus at the senior U.S. levels that this was central to the strategy undermined the continuity necessary for such efforts to succeed.[42] In its regional overtures, SRAP recognized the need for a political strategy to factor in neighboring countries that can be pivotal in tipping the scales toward war or peace. For example, satisfying Afghan Pashtun concerns provides less fertile ground for Pakistan, and securing Sunni equities in Iraq gives Gulf states less traction for anti-Shia efforts.

A political strategy will always have to account for the differing interests of the parties to a conflict. The mere fact that U.S. interests do not align with the host nation or antagonists' interests does not make political solutions impossible; indeed that is the business of diplomacy rightly understood, backed up not only by military force in a wartime context. Effective suasion very often requires conditionality, or the application of concrete sticks and carrots. Petraeus and Crocker practiced conditionality informally, but as Biddle notes, "in Afghanistan, by contrast, the West has been systematically unwilling to use conditionality as leverage for reform."[43]

Another possible reason for the blind spot or the lack of attention paid to political strategy is the military's discomfort with the "political lane." For example, the primary focus of the U.S. military in this regard in the early years in Iraq was to provide security for the political actors and to ensure sufficient security for the 2005 elections to be conducted. As time went on, important initiatives were launched to reconcile and reintegrate insurgent fighters, including those in detention. After a great deal of debate, however, in the past year the Army has embraced the notion of "engagement" and codified it as a seventh

---

[42]   Robinson (2008), pp. 250–260, outlines the elements and mechanisms for a political solution in Iraq, including a more federalized system that provides political representation, local security, equitable sharing of oil resources, and resolution of Kirkuk's status. See also Chandrasekaran (2012) for a description of efforts to share power and resources equitably in Afghanistan.

[43]   Stephen Biddle, "Afghanistan's Legacy: Emerging Lessons of an Ongoing War," *The Washington Quarterly*, Vol. 37, No. 2, 2014, p. 80.

warfighting function (along with movement and maneuver, fires, intelligence, sustainment, command and control [or mission command], and protection).[44] The function was added in order to enable the force to address this aspect of war, which it framed in this way: "How does the Army operate more effectively in the land domain while fully accounting for the human aspects of conflict and war by providing lethal and nonlethal capabilities to assess, shape, deter, and influence the decisions of security forces, governments, and people?"[45]

The argument is sometimes made that this type of interference is somehow illegitimate for the military because politics is a separate domain from war, or because the military should not engage in invasive political activities. In this view, military action should be limited to the exercise of physical violence rather than trying to change the political environment in foreign countries—an argument most often advanced by critics who oppose U.S. efforts to foster democracy in other countries.[46] This is why some scholars, like Gray and Weigley, have characterized the American way of war as "apolitical," because of its hesitancy to embrace political warfare and its tendency to view war as a technocratic application of systems, models, and technology. They argue that this view of warfare has serious weaknesses. Violence can never be so completely isolated from political considerations; it requires a political strategy to guide its use. More pragmatically, it is unclear why there should be such a firm dividing line between military action and political warfare. The act of making war is in itself highly invasive; as one workshop participant noted: "Why is it okay to massively impose our will on another country physically but not politically?"

Another impediment to understanding the centrality of the political dimension has been the counterinsurgency focus on building governing capacity. The U.S. Army Counterinsurgency Field Manual of 2006 stated that "[t]he primary objective of any COIN

---

[44] U.S. Department of the Army, *TRADOC Pam 525-8-5, U.S. Army Functional Concept for Engagement, 2014,* Washington, D.C.: Government Printing Office, February 2014.

[45] Department of the Army, *TRADOC Pam 525-8-5,* 2014, p. 7.

[46] See, for example, Anatol Lieven and John Hulsman, *Ethical Realism,* New York: Random House, 2009.

operation is to foster development of effective governance by a legitimate government."[47] This injunction was interpreted by both military and civilian implementers to mean building governing capacity in a technocratic sense, when the fundamental requirement was achieving a transition from violent to nonviolent competition. The distinction between technocratic skills and political solutions that work for the country in question is rarely made but critical to the strategic outcome. Biddle recently reached this conclusion: "What Afghanistan actually shows is that governance problems are about political *interests* at least as much as administrative capacity. Merely improving the capacity of government actors who seek to prey on their population makes things worse, not better. The United States needs a different approach to governance in counterinsurgency. . . . This includes the knowledge of how to balance security with a more political understanding of governance reform."[48]

The military might like to leave the political strategy entirely to the diplomats, but it is very difficult to conceive of a war-ending strategy without this dimension. Therefore, as Nadia Schadlow has argued, it is important to "consider the establishment of political and economic order as *a part of war itself*, the design and implementation of which requires both land forces—usually the Army—and an operationally minded diplomatic corps."[49] A RAND study on the war in Iraq reached a similar conclusion, noting that "wars do not end when major conflict ends. Wars emerge from an unsatisfactory set of political circumstances, and they end with the creation of new political circumstances that are more favorable to the victor."[50] Even when diplomats shoulder the main work of a war-ending negotiated settlement or other

---

[47] U.S. Department of the Army, *The U.S. Army-Marine Corps Counterinsurgency Field Manual, U.S. Army Field Manual No. 3-24, U.S. Marine Corps Warfighting Publication No. 3-33.5*, Chicago: University of Chicago Press, 2007, p. 37.

[48] Biddle, 2014, pp. 80, 81, 83–84.

[49] Nadia Schadlow, "War and the Art of Governance," *Parameters*, Vol. 33, No. 3, 2003, pp. 85–94; and Nadia Schadlow, "Competitive Engagement: Upgrading America's Influence," *Orbis*, Vol. 57, No. 4, 2013, pp. 501–515.

[50] Bensahel et al., 2008, p. 241.

elements of a political strategy, as they did, for example, in El Salvador in the 1990s, the military activities must be planned within the context of and in service to the political strategy's objectives.[51]

## 4. Technology Cannot Substitute for Expertise in History, Culture, and Languages Because of the Inherently Human and Uncertain Nature of War

As noted earlier, during the 1990s the military establishment and many defense intellectuals argued that the advent of networked computers and telecommunications had ushered in a "revolution in military affairs" (RMA) which would enable the United States to substitute high technology for manpower, and that the new forms of intelligence available to commanders would give them "information dominance."[52] This thinking was influenced by the 1991 Persian Gulf War in which the U.S. effectively leveraged relatively new technologies, including precision-guided munitions, global telecommunications, and unmanned aerial vehicles, to gain an unprecedented degree of precise information about the location and disposition of an opposing conventional military force and destroyed it accurately, thoroughly, and quickly. Some analysts saw the Persian Gulf War as a template for future conflicts and thus emphasized the development and procurement of high technology for future military planning. Some scholars viewed the early campaign in Afghanistan as ratifying this new model of warfare, in which very few U.S. personnel embedded with local allies, empowered by global telecommunications, U.S. airpower, and precision-guided munitions, are able to achieve military effects far out of proportion to their numbers.[53]

---

[51] Linda Robinson, "The End of El Salvador's War," *Survival*, Vol. 33, September/October 1991.

[52] For the debate on the revolution in military affairs, see Michael O'Hanlon, *Technological Change and the Future of Warfare*, Washington, D.C.: Brookings Institution Press, 2000; and Michael J. Mazarr, *The Revolution in Military Affairs: A Framework for Defense Planning*, Army War College, Strategic Studies Institute, Carlisle Barracks, Pa., 1994.

[53] Richard B. Andres, Craig Wills, and Thomas E. Griffith, Jr., "Winning with Allies: The Strategic Value of the Afghan Model," *International Security*, Vol. 30, No. 3, Winter 2005/06, pp. 124–160.

In the past decade, technological advances in intelligence, surveillance, and reconnaissance have indeed permitted fusion of massive volumes of data from electronic, signals, and imagery intelligence to help locate "high-value targets," greatly facilitating the counterterrorism mission. The benefits of technology are clear—including the ability to identify and geolocate targets, confirm battle damage assessments, support stand-off attacks and indirect fires, and enhance long-range coordination with allied and partner units—although they do not amount to the omniscient total information dominance that RMA promised. Furthermore, high technology seems to have had its greatest impact supporting kinetic operations.

High expectations for technology in the 1990s may have lessened the emphasis on developing understanding based on human, cultural, and social intelligence that commanders found so lacking on Afghanistan and Iraq.[54] Aside from exaggerated expectations of the benefits to be delivered by technology, the lack of investment in sociocultural and historical knowledge may also be attributable to skepticism regarding the ability to sufficiently master the complexities of other cultures and uncertainty about which cultures or regions should be prioritized. In the cold war, the choice was clear: Russian language, history, and culture. The conflicts of the past 13 years required detailed, local, nuanced information about tribal loyalties, local leaders, and regional histories (e.g., the record of land disputes, conflict over water, or tribal feuds) to support nonkinetic operations, including civil affairs, reconstruction, humanitarian relief, and political engagement. Without some degree of understanding of the countries in which U.S. forces were operating, they were hard pressed to carry out these activities successfully. The failure to embrace these activities as essential parts of war and peace-

---

[54] Christopher J. Lamb, James Douglas Orton, Michael C. Davies and Theodore T. Pikulsky, "The Way Ahead for Human Terrain Teams," *Joint Force Quarterly,* Vol. 70, No. 3, 2013, pp. 21–29. Douglas G. Vincent, *Being Human Beings: The Domains and a Human Realm.* Strategy Research Project, U.S. Army War College. March 2013. U.S. Department of Defense, *Report of the Defense Science Board Task Force on: Understanding Human Dynamics.* Washington, D.C.: Office of the Under Secretary of Defense for Acquisition, Technology, and Logistics, 2009. Anna Simmons, *21st Century Cultures of War: Advantage Them,* Carlisle, Pa.: Foreign Policy Research Institute, April 2013.

making, and therefore core functions for military and civilian entities, is ultimately due to the narrow American view of war.

When faced with the need to address this deficit, the military expended a great deal of effort in creating the knowledge needed for understanding the conflict. It did so in three ways: by developing it within the force, seeking to create or leverage interagency centers of knowledge, and tapping expertise outside the government to inform plans, execution, and assessment.[55] Attempts were made to expand military intelligence collection and analysis beyond its traditional enemy-oriented focus to a broader scope that included all the actors within the conflict arena.[56] Elsewhere within the military, the type of country expertise that was needed was limited to the small foreign area officer cadre and some elements of SOF. Civilian government experts were also tapped for their knowledge, but their numbers and deployability were limited in comparison to the demand. Social scientists were enlisted to provide expertise in Human Terrain Teams, but this concept proved difficult to implement, in part due to a shortage of academic experts qualified in the actual microcultures in which the military was operating. Over time, through military, governmental, and external efforts, the body of knowledge and the enterprise of sociocultural information-gathering and analysis became quite robust at multiple echelons, command headquarters, and combatant commands, as well as in a multitude of JIIM forums and virtual networks. The purpose of these multiple efforts was not to generate information, intelligence, or knowledge for academic purposes, but to gain sufficient and relevant understanding to craft and adapt strategy to the realities on the ground.

Just as important as the need to develop an understanding of the human, political, and sociocultural aspects of a conflict is the ability to place knowledge in historical context and use history as a guide to understanding. This is another truism without which policymak-

---

[55] Ben Connable, Walter L. Perry, Christopher Paul, K. Scott McMahon, Erin York, and Todd Nichols, "Geospatially-Focused Socio-Cultural Analysis at the U.S. Central Command's Afghanistan-Pakistan Center: A Review of the Human Terrain Analysis Branch (HTAB)," unpublished RAND Corporation research, 2012.

[56] Michael T. Flynn et al., "Fixing Intel: Making Intelligence Relevant in Afghanistan," Washington, D.C.: Center for a New American Security, January 4, 2010.

ers, strategists, and those charged with implementing strategy are lost. Without a collective and comprehensive understanding of what happened previously, there is little chance of developing an appropriate approach to new challenges. The tendency to rely on anecdote or personal experience in framing policy arguments was noted by a number of participants, and the only remedy is development of a historical mindset and a study of history that is full and contextual.[57]

The other critical use of knowledge is in the assessment function that is vital to adapting policy, strategy, and campaign plans over time, in response to developments and actions of the multiple players involved, as discussed in the second lesson and as prescribed in the Army's design method. This adaptive approach also recognizes that initial understanding will be imperfect. Gates noted that he considered periodic reviews to be vital to determine whether a policy was on track and to hold both the U.S. and partner governments to account. Moreover, Congress increasingly has made formal assessments a legislated requirement as part of its determination of whether to continue funding a given effort. For all these reasons a very large assessment enterprise has developed, and in the quest for increasingly refined and verifiable measures of effectiveness, some very complex methods have been developed.

Complexity does not always produce clarity, however, and very often—especially in war zones and developing countries—the desired data are unreliable or unobtainable. As a consumer and provider of such assessments, McMaster notes both the importance and the difficulty with assessments: "It is difficult to overstate the importance of constant reassessment. The nature of a conflict will continue to evolve because of continuous interaction with enemies and other destabilizing factors. Progress will never be linear, and there will have to be constant refinements and readjustments to even the best plans." But, he adds, "Commanders and senior civilian officials should be aware that overreliance on systems analysis can create an illusion of control and progress. Metrics often tell commanders and civilian officials how

---

[57]   Eliot A. Cohen, "The Historical Mind and Military Strategy," *Orbis*, Vol. 49, No. 4, Fall 2005, pp. 575–588.

they are executing their plan (e.g., money spent, numbers of indigenous forces trained and equipped, districts or provinces transferred to indigenous control), but fail to highlight logical disconnects. . . . An overreliance on metrics can lead to a tendency to develop short-term solutions for long-term problems and a focus on simplistic charts rather than on deliberate examinations of questions and issues critical to the war effort."[58] Metrics tend to characterize how programs and operations are doing (measures of performance), but not whether they are the right ones (measures of effectiveness). In other words, a comprehensive suite of metrics may tell a commander whether his strategy is being well executed, but not whether it is a successful one.

### 5. Interventions Should Not Be Conducted Without a Plan to Conduct Stability Operations, Capacity Building, Transition, and, If Necessary, Counterinsurgency

U.S. policymakers deliberately eschewed preparing for post-combat stabilization in both Iraq and Afghanistan, and the U.S. military did not deploy sufficient numbers of prepared forces to conduct stability operations as the need for them became evident, until a wrenching and prolonged review process late in both campaigns.[59] In Libya, the United States again chose to avoid undertaking a stabilization mission after removing the Gaddafi regime, opening the way for a climate of militia-dominated anarchy. The latter case failed to learn from the Iraq experience, in which the policy decision not to plan for robust stability operations led to the consequent failure by the U.S. military to have enough troops prepared and ready to do the mission.

For some, the failure to heed past experience in this regard almost defies comprehension, given voluminous evidence and scholarship suggesting that the need to consolidate the peace is a co-equal imperative

---

[58] McMaster, 2010, p. 103.

[59] Stability operations is an umbrella term for various military missions, tasks, and activities conducted outside the United States in coordination with other instruments of national power to maintain or reestablish a safe and secure environment and to provide essential governmental services, emergency infrastructure reconstruction, and humanitarian relief. (See Joint Publication 3-07, *Stability Operations*, 2011.)

of winning the war.[60] After participating in successive nation-building endeavors in Somalia, Haiti, Bosnia, Kosovo, Afghanistan, and Iraq, retired diplomat and scholar James Dobbins asked, "How . . . could the United States perform this mission so frequently yet do it so poorly?" His answer: "neither the American military nor any of the relevant civilian agencies had regarded post-conflict stabilization and reconstruction as a core function, to be adequately funded, regularly practiced, and routinely executed. . . . [The government] treated each mission as if it were the last such it would ever have to do."[61]

Despite attempts to mandate the planning, prioritization, and resourcing of these operations, the military has not historically embraced the missions, and the United States has often been reluctant to embrace the long-duration and sometimes large-scale operations that may be required. There is a well-established school of thought and U.S. tradition opposing intervention as imperialistic or unrealistic and embracing isolationism or a philosophy of limited engagement. But if the United States does decide to intervene, experience in the Balkans suggests that a sufficiently large footprint to conduct early stability operations, during the "golden hour" before any opposition can get organized, may prevent the need for a larger or longer deployment later.[62] Doing so requires military forces to prepare and train for policing operations in post-combat environments to prevent a breakdown in law and order. In some cases, indigenous forces may be able to perform

---

[60] Dobbins, James, John G. McGinn, Keith Crane, Seth G. Jones, Rollie Lal, Andrew Rathmell, Rachel Swanger, and Anga Timilsina, *America's Role in Nation-Building: From Germany to Iraq*, Santa Monica, Calif.: RAND Corporation, MR-1753-RC, 2003. Nora Bensahel, Olga Oliker, and Heather Peterson, *Improving Capacity for Stabilization and Reconstruction Operations,* Santa Monica, Calif.: RAND Corporation, MG-852-OSD, 2009. Beth Cole, "Guiding Principles for Stabilization and Reconstruction," United States Army Peacekeeping and Stability Operations Institute, United States Institute of Peace, 2009.

[61] James Dobbins, "Retaining the Lessons of Nationbuilding," in *Commanding Heights: Strategic Lessons from Complex Operations*, Washington, D.C.: NDU Press, 2010, p. 65.

[62] Dobbins et al., 2003. See Dobbins, James, Seth G. Jones, Benjamin Runkle, and Siddharth Mohandas, *Occupying Iraq: A History of the Coalition Provisional Authority*, Santa Monica, Calif.: RAND Corporation, MG-847-CC, 2009, for discussion of the lack of a deliberate program for disarming, demobilizing, and reintegrating the former Iraqi security force members, pp. xxiii, 501–580.

or assist with such operations, but a careful evaluation of their loyalty and competence is required.

Experience also suggests that building partner capacity to get local security forces in the lead as quickly as possible should be a higher priority. In both Iraq and Afghanistan, building indigenous security forces took second place to combat operations in terms of the urgency and resources devoted to the task, with particularly severe deficits in building adequate police, logistic and other enablers, and headquarters staff and institutions.[63] In Afghanistan, there was no reexamination of security force assistance and security sector reform—the key to the U.S. strategy there—despite drastic changes in the security environment between 2002 and 2007.[64]

In an attempt to enshrine these imperatives, U.S. military doctrine has been revised to adopt a numerical phasing concept that implies that Phase III (major combat operations) should be followed by Phase IV (stability operations), and then Phase V (transition to civilian control and a resumption of normal government functions). Conrad Crane observed that the military phasing construct can be problematic as elements of phases may not divide neatly, and planning, training, and resourcing of the phases must, to some extent, be done simultaneously. He wrote that, "Even the concept of having separate phases during a campaign might be worth rethinking because the con-

---

[63] For example, there were only 4,000 advisers to mentor the 300,000-strong nascent Iraq security forces as of 2006 (Robinson, 2008, p. 95). RAND has published numerous studies documenting the gaps and efforts to improve the U.S. model and ability to conduct stability operations and build security forces, including Thomas S. Szayna, Derek Eaton, and Amy Richardson, *Preparing the Army for Stability Operations: Doctrinal and Interagency Issues*, Santa Monica, Calif.: RAND Corporation, MG-646-A, 2007; Jefferson P. Marquis, Jennifer D. P. Moroney, Justin Beck, Derek Eaton, Scott Hiromoto, David R. Howell, Janet Lewis, Charlotte Lynch, Michael J. Neumann, and Cathryn Quantic Thurston, *Developing an Army Strategy for Building Partner Capacity for Stability Operations*, Santa Monica, Calif.: RAND Corporation, MG-942-A, 2010; and Christopher Paul, Colin P. Clarke, Beth Grill, Stephanie Young, Jennifer D. P. Moroney, Joe Hogler, and Christine Leah, *What Works Best When Building Partner Capacity and Under What Circumstances?* Santa Monica, Calif.: RAND Corporation, MG-1253/1-OSD, 2013, as well as other country-specific studies cited elsewhere in this report.

[64] Terrence K. Kelly, Nora Bensahel, and Olga Oliker, *Security Force Assistance in Afghanistan: Identifying Lessons for Future Efforts*, Santa Monica, Calif.: RAND Corporation, MG-1066-A, 2011.

struct can stovepipe planning and hamper the holistic vision necessary to properly link combat to the end state that accomplishes national political objectives."[65] Regardless of the utility of numbered phases, the essential lesson is that interventions are rarely successful if these critical post-combat operations are not planned, resourced, and conducted.

Transition to civil authority, or Phase V operations, is equally important. If Phase IV was rocky in Iraq, and capacity-building deficient in Iraq, Afghanistan, and Libya, Phase V in Iraq did not go smoothly either (that phase is now approaching in Afghanistan). As RAND research has noted, "The planning for the withdrawal of U.S. troops from Iraq and transition to civilian lead suffered from a "gap between established goals and the time and resources necessary to achieve them," leaving civilian and military personnel facing the "challenge of seeking to achieve overly optimistic strategic and policy goals with insufficient resources."[66] The U.S. general in charge of leading the U.S. embassy's office of security cooperation detailed the numerous ways in which he could not carry out his intended functions due to "outdated statutory authorities not designed for today's operating environment." including training and advising, provision of institutional support, security assistance to the Iraqi federal police, and internal security support.[67] The failures were not only planning failures or

---

[65]   Conrad Crane, "Phase IV Operations: Where Wars Are Really Won," *Military Review*, May–June 2005, p. 11.

[66]   Rick Brennan, Jr., Charles P. Ries, Larry Hanauer, Ben Connable, Terrence K. Kelly, Michael J. McNerney, Stephanie Young, Jason H. Campbell, and K. Scott McMahon, "Smooth Transitions? Lessons Learned from Transferring U.S. Military Responsibilities to Civilian Authorities in Iraq," Santa Monica, Calif.: RAND Corporation, RB-9749-USFI, 2013, p. 4. This brief makes numerous recommendations for improving transition and Phase V execution, including "Military planners should make institution-building a priority to ensure that the progress made through training, advising, and assisting will be sustained after the transition" (p. 7). This is equally applicable to the return on investment made in building the Afghan security forces post-2014. The full report is Rick Brennan, Jr., Charles P. Ries, Larry Hanauer, Ben Connable, Terrence K. Kelly, Michael J. McNerney, Stephanie Young, Jason H. Campbell, and K. Scott McMahon, *Ending the U.S. War in Iraq*, Santa Monica, Calif.: RAND Corporation, RR-232-USFI, 2013.

[67]   Robert L. Caslen, Jr., et al., "Security Cooperation Doctrine and Authorities: Closing the Gaps," *Joint Force Quarterly*, Vol. 71, No. 4, p. 74.

policy gaps such as needed authorities, but also, as one workshop participant commented, "the lack of thought about the political strategy of withdrawal. What would be the impact of withdrawal on Maliki and others? We were more focused on technocratic issues, how to transfer responsibility from the military to the embassy, and the military never thought about the political side."

In addition to revising military doctrine, the U.S. government took some steps to institutionalize the needed approach. In December 2005, the Bush administration issued National Security Presidential Directive 44 (NSPD-44), "Management of Interagency Efforts Concerning Reconstruction and Stabilization." NSPD-44 designated the Secretary of State as the lead agent for coordinating interagency planning efforts. By doing so it implied that reconstruction and stabilization operations were primarily a task for civilian agencies: It tasked the Secretary to "coordinate" with the Secretary of Defense "to ensure harmonization with any planned or ongoing U.S. military operations," which merely stated, but did not resolve, the problem of civil-military coordination in complex operations.

The current policy does not appear to provide adequate assurance that JIIM partners will achieve the necessary integration of effort or resourcing for their requirements. Despite the issuing of directives (there are also DoD directives relating to Stability Operations and Irregular Warfare) and doctrine (including novel experiments in interagency doctrine for both stability and reconstruction operations and counterinsurgency), there is no guarantee that even the progress in approach made over the past 13 years will be lasting. In Dobbins' view, "much still needs to be done if the current level of expertise is not to degrade again after the immediate crises recede. Forestalling such a regression will require the establishment, *by legislation,* of an enduring division of labor between the White House, the Department of State, DoD, and USAID. There must be an allocation of responsibilities that cannot be lightly altered by each passing administration, for no agency will invest in activities it may not long need to carry out."[68]

---

[68] Dobbins, 2010, p. 71 (emphasis added).

Drawing on lessons from the Balkans as well as the recent wars, several workshop participants suggested that robustly resourced stability operations initiated soon after the intervention might have forestalled much of the later chaos. It is not possible to prove the assertion but it is useful nonetheless to ask whether robustly implemented stability operations, including the critical task of policing to maintain public order, followed quickly by a comprehensive effort to build adequate security forces, would have largely secured the peace. Todd Greentree, a diplomat who served several tours in Afghanistan, argues that, in that case, "[t]he signal error was failure to develop the ANSF [Afghan Security Forces] while the Taliban and al Qaeda were at their weakest. Doing so early on would have made it possible for the ANSF to maintain internal security while remaining a modest and sustainable size."[69] A RAND study also assessed that for indigenous security forces to be successful, they need to exist in the right quantity, of adequate quality and with the right loyalty.[70] The analysis of this project team suggests that stability operations and building indigenous security force capacity were necessary but not sufficient elements of a successful approach: The additional needed component was a political strategy that provided a place in the new political order for the major political factions of each country, on terms that the potential spoilers in the region (e.g., Pakistan, Iran, Saudi Arabia) would accept.

In the event, U.S. actions, inaction, and other circumstances led to growing insurgencies in both Iraq and Afghanistan. The U.S. government faced a choice between abandoning the enterprise or mounting a large-scale counterinsurgency effort. It embarked on the latter path and devoted troops, effort, and resources to quelling the combination of insurgency, terrorism, and sectarian war between 2003 and 2011 when it departed Iraq and 2014 as it transitions or departs Afghanistan.[71]

---

[69]    Todd R. Greentree, "Lessons from Limited Wars: A War Examined, Afghanistan," *Parameters,* Vol. 43, No. 3, 2013, p. 93.

[70]    Kelly, Bensahel, and Oliker, 2011.

[71]    As of this writing, the completed but unsigned Bilateral Security Agreement and Status of Forces Agreement have yet to be implemented to pave the way for a small-scale advisory

Some critics have argued that the experience in Iraq and Afghanistan demonstrated the futility of counterinsurgency and stability operations.[72] They correctly argue that counterinsurgency is not a strategy, but the charge that COIN is overly "soft" overlooks the degree to which enemy-centric, kinetic operations were part of the effort. The much-used "hearts and minds" phrase obscures the fact that the central objective of population-centric counterinsurgency tactics is to separate the population from the adversary.[73] Perhaps the most important lesson to learn from these war years is that counterinsurgency is a time- and resource-intensive endeavor, at least as recently employed with the United States acting as the primary counterinsurgent. The United States must therefore ensure that its interests warrant that expenditure of effort, which includes weighing the sunk costs of an endeavor as well as the loss of regional or international clout as the result of a failed war or abandonment of allies. Many U.S. allies, most notably NATO, have invested significant effort and prestige in these wars as well. Because the United States encountered unexpected challenges and achieved suboptimal outcomes in Iraq and Afghanistan, some are inclined to advise that the United States simply not fight counterinsurgencies or undertake stability operations in the future.

Other alternatives are preferable, including the option of supporting indigenous counterinsurgency efforts, called foreign internal defense (FID) in U.S. doctrine, which is discussed in the next lesson. In that case the United States provides assistance, which could possibly include combat advisers, but the weight of the effort is borne by the country afflicted by the insurgency. This was not an option by the time the war in Iraq was a full-blown sectarian conflict, since the Iraqi government was a major antagonist in a fight against some of its own

---

and counterterrorism mission by U.S. and some NATO forces after 2014.

[72] Gian Gentile, *Wrong Turn: America's Deadly Embrace of Counterinsurgency*, New York: The New Press, 2013; Douglas Porch, *Counterinsurgency: Exposing the Myths of the New Way of War*, New York: Cambridge University Press, 2013.

[73] Raphael S. Cohen, "Just How Important Are 'Hearts and Minds' Anyway? Counterinsurgency Goes to the Polls," *Journal of Strategic Studies*, ahead of print, 2014, pp. 1–28.

constituents. The function the coalition troops played was as much one of peace enforcement as counterinsurgency.

The counterinsurgency tactics employed during the surge did decrease levels of violence, but the fact that they were not carried out in the larger context of a concerted, sustained implementation of a political strategy makes it difficult to settle the argument as to the tactics' effectiveness. Many caricatured arguments have been formulated suggesting that the U.S. actions had no effect and that other actors' actions were entirely responsible for the demonstrated decline in violence achieved in both "surges."[74] Biddle has examined both sides and most recently concluded that "the Iraq surge did suggest that COIN was not impossible," and "the Afghan experience shows that current U.S. methods *can* return threatened districts to government control, when conducted with the necessary time and resources." After reviewing the outcome of COIN efforts in different districts and provinces in Afghanistan he concluded: "The commonplace narrative that Afghanistan shows how COIN is impossible is thus overstated. What experience to date suggests is that it can work—but only where counterinsurgents invest the lives, forces, and time needed." [75] He also concluded that the surge in Iraq played a key role creating a "synergistic reaction" with the Anbar Awakening that "created something new that neither could have achieved alone."[76]

COIN was difficult and costly in Iraq and Afghanistan in part because the U.S. military had not trained or prepared for such missions, and in part because it was inadequately tethered to a realistic political strategy. Criticizing COIN and stability operations for being too costly and difficult thus becomes a self-fulfilling prophecy: By refusing to train and prepare for such operations because of their cost and risk, the military only makes it more likely that such operations

---

[74]   David Ucko, "Critics Gone Wild: Counterinsurgency as the Root of All Evil," *Small Wars & Insurgencies*, Vol. 25, No. 1, 2014, pp. 161–179.

[75]   Stephen D. Biddle, "Afghanistan's Legacy: Emerging Lessons of an Ongoing War," *The Washington Quarterly*, Vol. 37, No. 2, 2014, pp. 75–76, 78.

[76]   Stephen Biddle, Jeffrey A. Friedman, and Jacob N. Shapiro, "Testing the Surge: Why Did Violence Decline in Iraq in 2007?" *International Security*, Vol. 37, No. 1, 2012, pp. 10–11.

will be more costly and risky than necessary. If military and civilian agencies retain their hard-won expertise in counterinsurgency, stability operations, and building partner capacity, they will help keep the costs of such operations down and increase their likelihood of success. Nonetheless, the option of engaging in a large-scale counterinsurgency should be considered a distant second when other alternatives are available, as will be discussed in the next lesson.

### 6. Shaping, Influence, and Unconventional Operations May Be Cost-Effective Ways of Addressing Conflict That Obviate the Need for Larger, Costlier Interventions

Proactive, preventive engagement is considered by some to be an obvious and even prudent middle-ground alternative between inaction and full-scale intervention. For others, it represents a dubious proposition that most often leads down a slippery slope to extended commitments and unsatisfactory results. The defense strategic guidance and the Quadrennial Defense Review embrace the notion that building partner capacity and using partners can be cost-efficient ways of securing U.S. interests, although the actual resources devoted to this middle path are quite modest. The argument is that if U.S. government invested more in such efforts, it might sufficiently mitigate the threat or conflict to avoid the need for Phase III interventions. Workshop participants noted that the Phase 0 suite of prevention, shaping, and influence operations has received far less attention and emphasis, particularly compared to Phases III, IV, and V. As one participant noted, "Everything I hear from the military suggests that we don't have a theory of victory in Phase 0. We have to do a lot better job at conceiving and implementing Phase 0 and I [the latter is deterrence operations]. We don't have a clear idea of what needs to be done or how we will do it." In part this is due to an inadequate conceptualization of how SOF can be employed both separately and in conjunction with other forces to achieve strategic effects. They have historically been used as a tactical force, but as their role has grown so has the need for operational art and operational level command to accommodate and exploit their unique advantages. The use of SOF has grown dramatically in the past 13 years in all phases of military operations, but their use has not yet been translated into a new way of war.

Critics of this approach have pointed out significant challenges to effective preventive engagement. They argue that the United States is at Phase 0 or I in almost all countries in the world; it cannot invest more everywhere. Policymakers need to set priorities as to where to engage based partly on which countries are most at risk, but the intelligence community does not have a good track record of projecting where or when conflict will next erupt, much less specific conflicts that might threaten U.S. interests.[77] And even if it could, they argue, the United States has rarely demonstrated the ability to develop coordinated, proactive, far-sighted initiatives abroad.[78] Investing more in preventive engagement might thus be a well-intentioned but impracticable option that leads the United States to make needless, expensive, and ultimately ineffectual investments in strategically secondary theaters.[79]

This critique may overstate the difficulties of preventive engagement. While there are nearly 200 states in the world, there are far fewer that are both vital to U.S. interests and at risk of conflict. Policymakers do need to identify those states in key regions that would benefit from greater U.S. engagement. Another criticism is one of efficacy—i.e., that it does not produce the desired results. There is a need to approach this course of action with a dose of realism rather than wildly aspirational objectives. In many cases, the shaping and influence operations may be a way to achieve a "good enough" outcome at reasonable cost, though they may often require a long timeline to produce the desired results.

---

[77] The Political Instability Task Force found that the most reliable single indicator of impending state failure, civil conflict, ethnic cleansing, or war crimes was, interestingly, infant mortality—likely a proxy indicator that captured a host of failures in governance, legitimacy, and capacity across the board. The indicator, however, can be difficult to collect in real time and thus lacks predictive value.

[78] Richard K. Betts, "Is Strategy an Illusion?" *International Security,* Vol. 25, No. 2, 2000, pp. 5–50.

[79] This tends to be the line of reasoning among advocates of a "restrained" or "limited" U.S. role in the world. See, for example, Barry R. Posen, "The Case for Restraint," *The American Interest,* Vol. 3, No. 1, 2007; Patrick M. Cronin, *Restraint: Recalibrating American Strategy,* Washington, D.C.: Center for a New American Security, 2010; Stephen M. Walt, *Taming American Power: The Global Response to U.S. Primacy,* New York: W. W. Norton, 2005; Michael Mandelbaum, *The Frugal Superpower: America's Global Leadership in a Cash-Strapped Era,* New York: PublicAffairs, 2011.

Broadly, policymakers should recognize and be prepared to adjudicate some of the typical trade-offs.[80] For example, they can choose to work through existing, flawed institutions at lower cost but also with less latitude for ambitious reforms, or they can try to replace existing institutions with new ones at higher cost but potentially greater long-term benefit. More generally, since prevention, shaping, and influence involve working with "what is"—i.e., governments, security forces, and informal partners—there will almost always be such serious issues as competence, divergent interests, corruption and patronage, human rights abuses, and anti-democratic tendencies. Those issues are not incidental. At the latter end of the spectrum of risks, the assistance, shaping, and influence could turn out to be providing the wherewithal for a coup to occur—or sustaining support for a dictator. The United States has done both of these throughout its history.

To become a more viable default option, shaping, influence, and unconventional operations must demonstrate efficacy and improve the identification and management of such risks. There is thus a need for further development of a model that will help policymakers apply the Phase 0 toolkit to achieve national security objectives. Some research has been done on the conditions under which assistance to partners is likely to produce the desired results, but more work is needed to identify the necessary and sufficient conditions and apply them to the current gamut of threats to determine where they are most likely to bear fruit.[81] The first requirement to develop a systematic approach is to understand the full range of activities that may be undertaken as part of a concerted campaign that does not involve major combat by U.S. forces.

---

[80] Paul D. Miller, "Armed State Building: Confronting State Failure, 1898–2012," Ithaca, N.Y.: Cornell University Press, 2013.

[81] Stephen Watts, Jason H. Campbell, Patrick B. Johnston, Sameer Lalwani, and Sarah H. Bana, *Countering Others' Insurgencies: Understanding U.S. Small-Footprint Interventions in Local Context*, Santa Monica, Calif.: RAND Corporation, RR-513-SRF, 2014; and Thomas S. Szayna, Derek Eaton, Stephen Watts, Joshua Klimas, James T. Quinlivan, and James C. Crowley, *Assessing Alternatives for Full Spectrum Operations and Security Force Assistance: Specialized vs. Multipurpose Army Forces*, unpublished RAND Corporation research, 2013.

In conflict environments that are hostile, denied, or politically sensitive, SOF are likely to carry out a preponderance of these activities to shape, influence, and support partners (which may be government forces or informal groups, such as tribes or militias) because they are selected, trained, and equipped to operate in very small formations in a low-visibility or clandestine manner. For example, FID and unconventional warfare (UW) are two operations that can be carried out entirely in Phase 0.[82] FID is the provision of military and civilian assistance to a government under threat, in accordance with that government's own "internal defense and development" plan. The U.S. operation in the Philippines is an example of an effective foreign internal defense mission. After the September 11, 2001, attacks, the U.S. deployed a small contingent of SOF to the Philippines for Operation Enduring Freedom—Philippines (OEF-P) to help train and advise Philippine forces combating the Abu Sayyaf Group (ASG) and Jemmah Islamiyya (JI), two terrorist groups with links to al Qaeda. The groups had not attacked the United States and were comparatively smaller and posed less of a threat to the Philippines than al Qaeda posed to Afghanistan or Pakistan, yet the United States nonetheless devoted some resources to combating them and training partner security forces. The U.S intervention was smaller by orders of magnitude than the deployments to Iraq or Afghanistan, never totaling more than about 600 troops. And it was successful: The two groups, small to begin with, were even more constrained and limited, and Philippine forces were more capable of managing the threat from them, after a decade of sustained but relatively small U.S. investment. The intervention benefited from its proactive nature: The United States did not wait until the Philippine government was near collapse or ASG on the brink of overrunning the

---

[82] FID is defined as "participation by civilian and military agencies of a government in any of the action programs taken by another government or other designated organization to free and protect its society from subversion, lawlessness, insurgency, terrorism, and other threats to its security" (JP 3-22). Unconventional warfare is another method of assisting a friendly force, in this case to coerce or remove a government or occupying power. The doctrinal definition is "activities to enable a resistance or insurgency to coerce, disrupt or overthrow a government or occupying power through and with an underground, auxiliary, and guerrilla force in a denied area" (JP 1-02).

government. Because the situation was not dire, it was also relatively less costly to deal with.

Other influence operations and security cooperation activities can be part of a noncombat approach to achieving national security objectives. Civil affairs units are the most-deployed units in the joint force, and active-duty civil affairs are a core part of almost all special operations efforts. Similarly, military information support operations (formerly known as psychological operations) are employed in virtually all special operations endeavors, as well as in support of embassy and public diplomacy efforts. They represent a powerful if underdeveloped capability for achieving nonlethal effects. "Phase 0" operations are rarely conducted by SOF alone. Conventional forces and interagency and multinational partners are frequently involved, such as in the effort to stabilize and support a new government in Somalia under the United Nations–sanctioned African Union peace enforcement mission AMISOM. Security force assistance and partner capacity building are often discussed as activities with no end state, but in fact they should be construed as operations that achieve effects through strengthening and supporting another country. Influence operations and shaping operations can also include counter threat finance, cyber operations, and many other nonlethal activities. All of these should be considered tools that can be assembled in innovative ways and applied with sufficient foresight and duration to achieve lasting effect without resort to major combat operations.[83]

If the usage of numerical phases for military operations is problematic in general, the term "Phase 0" is a particular obstacle in conveying the strategic potential of the shaping and influence operations. The term "Phase 0," which stems from the major conventional war paradigm, suggests that there is an ineluctable move from zero to a Phase III major combat operation. But instead this constellation of shaping, influence, and partnered activities may be envisioned and

---

[83] See Dan Madden, Bruce R. Nardulli, Dick Hoffmann, Michael Johnson, Fred T. Krawchuk, John E. Peters, Linda Robinson, and Abby Doll, *Toward Operational Art in Special Warfare*, Santa Monica, Calif.: RAND Corporation, forthcoming, for a discussion of how Phase 0 operations can be placed in a campaign context.

conducted as a campaign that aims to achieve the stated objectives without a large-scale combat operation by U.S. forces. This could be viewed as a first-resort option for achieving U.S. objectives. This alternative paradigm of shaping, prevention, and influence seeks to achieve effects and thus avoid the need for a large commitment of U.S. forces.

Phase 0 is also unhelpful in conveying the spectrum of options available under this approach. The foreign internal defense mission often involves combat, but the combat is undertaken primarily by the host nation forces, with advisory support from special operations and conventional forces that operate under a variety of rules of engagement in keeping with the U.S. and host nation agreement. The unconventional warfare mission is another support mission that provides support to indigenous forces that are fighting an occupation or seeking to overthrow a regime; this was used to topple the Taliban and might be used to coerce or disrupt the Syrian regime or recent Russian incursions in Europe. FID and UW may involve combat advisers, but the main effort is carried by the indigenous force.

The United States has opportunities to invest in shaping and influence operations around the world, including in Yemen and North Africa. Compared to the trillion dollar expenditures in Iraq and Afghanistan, preventive engagement costs less money, requires fewer troops, and incurs less risk than larger, conventional operations. The Philippines, of course, was a comparatively small threat and not too problematic an ally. While working with partners often is highly problematic, tedious, and time-consuming, it may be the best option considering the downsides of all the rest.

A strategy that seeks to avoid major combat operations and long, costly entanglements such as Iraq and Afghanistan would place increasing emphasis on what the joint force and its interagency, intergovernmental, and multinational partners can do in Phase 0 to address threats to U.S. interests. Applying such an approach would require in many cases a forward leaning proactive stance on the part of policymakers, as well as strategic patience for some of these initiatives to bear fruit. Yet in the long term the benefits of these investments could last for years, and the costs could be significantly less than the endeavors of the past 13 years. The administration, in its proposed counterterrorism partnership fund, appears to have in mind such an approach as a major feature of its evolving counterterrorism strategy, although the programmatic details and model (including criteria for supporting states, scope of assistance, and metrics and conditions for execution and assessment) have not been provided in the initial documents accompanying the funding request.[84]

## 7. The Joint Force Requires Nonmilitary and Multinational Partners, as Well as Structures for Coordinated Implementation

The joint force has recognized in doctrine and in practice that it needs interagency, intergovernmental, and multinational partners, but adequate funding and practices to prepare those partners is lacking. In addition, no practice currently mandates coordinated implementation under a unified command structure, leaving unity of effort to be achieved by voluntary means alone. Because of the lack of relevant expertise within its own ranks, the U.S. military realized that in many operations, particularly Phases 0, IV, and V, it could not accomplish the mission alone. It required civilian expertise from a host of agencies, and it benefited from the expertise and political support of multinational coalitions. The need for integration between civilians and the military is not limited to the making of policy and strategy; it is also a necessary part of policy implementation. Yet the United States cannot impose unity of command by military fiat across services, agencies, and coalitions, creating the challenge of integrating and coordinating efforts.[85]

---

[84]  See transcript of President Obama's speech at West Point, New York, on May 29, 2014; Under Secretary of Defense (Comptroller), "FY2015 Counterterrorism Partnerships Fund and the European Reassurance Initiative," undated; and Under Secretary of Defense (Comptroller), "FY2015 DoD Overseas Contingency Operations (OCO) Budget Amendment," undated.

[85]  Mick Ryan, "After Afghanistan: A Small Army and the Strategic Employment of Land Power," *Security Challenges*, forthcoming; Richard B. Andres, Craig Wills, and Thomas Griffith Jr., "Winning with Allies: The Strategic Value of the Afghan Model," *International Security*, Vol. 30, No. 3, Winter 2005/06, pp. 124–160; Michael Spirtas, Jennifer D. P. Moroney, Harry J. Thie, Joe Hogler, and Durell Young, *Department of Defense Training for Operations with Interagency, Multinational, and Coalition Partners*, Santa Monica, Calif.: RAND Corporation, MG-707-OSD, 2008; Szayna et al., 2009.

To take the issue of integrating structures first, the U.S. government's record in the latter years of the wars in Iraq and Afghanistan suggests that neither the Iraq nor the Afghanistan interagency structure succeeded in establishing an enduring framework for integrated civilian-military policy management. For example, the SRAP and the White House were in frequent conflict over both Afghanistan and Pakistan policy implementation, primarily because the White House preferred a minimalist solution while SRAP was seeking to promote stability through institution-building. And in Kabul, the International Security Assistance Force (ISAF) command and the U.S. embassy were at odds, particularly during the tenures of Gen. McChrystal and Amb. Karl Eikenberry. A U.S. civilian-military integrated implementation plan was finally forged between the two entities, but it critically left out the rest of the ISAF coalition.

Among the many experiments in civil-military interagency partnerships, the Provincial Reconstruction Teams (PRTs) were the best known and most widely employed in both Iraq and Afghanistan. Early in the war in Afghanistan the U.S. military developed the concept of PRTs, which were small civilian-military units led by military officers. They were in some cases manned by civilian reconstruction experts from the Departments of State, Agriculture, or Justice and the U.S. Agency for International Development, as well as NATO ally civilians. Their role was to foster governance and development in key areas. PRTs were later employed in Iraq under a different model; they were civilian-led and reported to the embassy rather than the regional military commands. The integrated civilian-military concept drew inspiration from the CORDS program in Vietnam with a crucial difference in their respective command structures.[86] CORDS established a civilian chain of command operating under the overall military command.

---

[86] The CORDS program under Ambassador Robert Komer was directly subordinate to the Military Assistance Command in Vietnam (MACV), the first and only time such an arrangement has been tried in U.S. history. Komer exercised authority over military subordinates and appointed military heads to each provincial CORDS team with a civilian deputy, or a civilian head with a military deputy. By contrast, civilian PRT personnel were never under the formal command of military officers, nor vice versa. See Thomas W. Scoville, *Reor-*

Two other experiments are worth noting. NATO established a Senior Civilian Representative as a counterpart to the ISAF NATO military command, and as a counterbalance to the U.S. ambassador's inevitably strong sway. The additional voice in Kabul made for some friction, but it was an appropriate way to give greater weight to both coalition and civilian perspectives on strategic-level issues. The U.S. State Department created five "Regional Platforms" (RPs) headed by a senior foreign service officer to serve as the "two-star" civilian counterpart to ISAF's five Regional Commands (RCs), in order to represent the civilian view and create a combined approach at the all-important regional level (where Afghan political dynamics were most potent). This was only as effective as the partnerships forged by the RP and RC commands, and by the RC command's commitment to the noncombat aspects of the plan.[87]

Given the decision not to implement the Vietnam-era CORDS model or to recreate the military governors who oversaw the reconstruction of West Germany and Japan, either of which would have imposed unity of command, the alternative in Iraq and Afghanistan was to seek "unity of effort." Similar coordination is needed in non-war theaters to achieve unity of effort. Under the National Security Decision Directive 38 (NSDD-38), the U.S. ambassador exercises authority over all civilian and military personnel assigned to the embassy, but the geographic combatant command retains authority over military personnel assigned to it under Title 10.[88]

In addition to difficulties in coordinated planning and implementation, there is a chronic shortage of civilian capability and capac-

---

*ganizing for Pacification Support*, Washington, D.C.: Center of Military History, U.S. Army, 1982, Chapter 5.

[87] See Chandrasekaran (2012) for the unhappy experience of an RC-South commander who actively sought close civil-military collaboration but found his counterpart so defensive that the latter put up a locked fence between the two compounds.

[88] The State Department website states that "The National Security Decision Directive (NSDD) 38 dated June 2, 1982, gives the Chief of Mission (COM) control of the size, composition, and mandate of overseas full-time mission staffing for all U.S. Government agencies."

ity to participate in stabilization and reconstruction missions.[89] The record of the past 13 years shows that too few civilians appropriately trained or qualified for the mission were provided in both Iraq and Afghanistan. In an effort to address the deficits in civilian capacity and capability, the State Department formed its Office of the Coordinator for Reconstruction and Stabilization—later upgraded to the Bureau of Conflict and Stabilization Operations (CSO)—to execute its responsibilities under NSPD-44.[90] Both the funding and the political will to establish a standing deployable civilian and police capacity for reconstruction and stabilization have been lacking, however, in part due to reluctance to engage in large-scale operations. The CSO's Civilian Response Corps (CRC) was originally envisioned as comprising 4,250 active, standby, and reserve deployable civilians, but it is now largely defunct, with only a rolodex of people who might be persuaded to deploy. USAID's model relies heavily on contracted "implementing partners," though it has reconstituted its component of the CRC as its own Crisis Surge Support Staff (CS3).[91] In Iraq and Afghanistan, the State Department addressed the capacity gap by hiring temporary "3161" personnel to serve on PRTs (and the local-level District Support Teams). Workshop participants noted that even with the drawdowns in Iraq and Afghanistan, the current demand to support humanitarian and stability missions in Syria and elsewhere has strained the capacity of the civilian interagency community.

---

[89]   See, for example, Hans Binnendijk and Patrick M. Cronin, *Civilian Surge: Key to Complex Operations*, National Defense University, Washington, D.C., Center for Technology and National Security Policy, 2009; and American Academy of Diplomacy, *Forging a 21st Century Diplomatic Service for the United States Through Professional Education and Training*, Washington, D.C.: The Henry L. Stimson Center, 2011.

[90]   Nina Serafino, "Peacekeeping/Stabilization and Conflict Transitions: Background and Congressional Action on the Civilian Response/Reserve Corps and other Civilian Stabilization and Reconstruction Capabilities," Congressional Research Service, October 2, 2012; National Security Presidential Directive 44 (NSPD-44), *Management of Interagency Efforts Concerning Reconstruction and Stabilization*, Washington, D.C., December 7, 2005; Presidential Decision Directive 56 (PDD-56), *Managing Complex Contingency Operations*, White Paper, May 1997.

[91]   Dobbins et al., 2003; Bensahel, Oliker, and Peterson, 2003.

A related and limiting issue that inhibits civilian effectiveness is the force protection rules under which civilians are able to perform their duties. In some cases civilians were critically limited in conducting their missions in hostile environments. The U.S. State Department, through each embassy's Regional Security Officer, imposes strict force protection requirements on personnel assigned to the embassy. In many cases this inhibited their ability to do their jobs and work alongside military personnel, particularly outside the capital. For example, as of 2011, U.S. civilians who were not assigned to a forward operating base in Afghanistan were not permitted to stay overnight there. If their mission in the field could not be conducted in one day, they were not able to leave Kabul. In addition, the majority of civilian personnel were assigned to Kabul throughout the "civilian surge" rather than deployed to the areas of conflict, with most drawn back to the embassy by 2012. The requirements to travel in armored convoys of a certain type also inhibited their movements because they had limited force protection resources.

The record of multinational coalition performance is far more positive, although the speed of coalition decisionmaking is, at least in the case of NATO coalitions, destined to remain the speed of the slowest national decisionmaking. NATO and NATO countries are not the only, but are by far the most important, coalition partners of the United States over the past 13 years, along with Australia, which has participated in both Iraq and Afghanistan in relatively substantial numbers. As with many other partners who joined both coalitions, the experience provided the opportunity to learn coalition operations, contribute to a common security objective, and earn potential compensating support for its own needs and interests.

The U.S. government has benefited diplomatically from the international support that coalitions provide. The U.S. military is tasked with fighting and winning the nation's wars, and while acting alone is at times required, the United States prefers to operate with other nations and entities in coalitions of the willing. Virtually all U.S. military operations have been part of a multinational coalition since 1989, many in partnership with NATO, including operations in Bosnia,

Serbia and Kosovo, Afghanistan, Libya, Pakistan, the Red Sea region, and Turkey.[92]

The *Decade of War* study did not specifically assess coalition operations, but numerous reports have examined NATO and U.S. military experiences with the goal of improving operational integration. What follows is a brief survey of the principal observations common to these reports. Specifically, NATO has conducted its own studies and produced several "lessons learned" papers, and earlier RAND reports have produced recommendations that are still relevant today.

NATO member militaries vary in size, structure, and capability, ranging from highly sophisticated militaries that are fully integrated into U.S. planning and operations to those that cannot function without tremendous U.S. assistance—most partners lack the ability to deploy and conduct sustained operations outside of their territory without considerable U.S. support. Command and control (C2) of NATO operations has remained difficult, with multiple command structures across the operational domain hindering operations. To mitigate issues of interoperability within the NATO structure, early and continuous planning to incorporate coalition support requirements is recommended.[93] Additionally, liaisons and liaison teams are necessary to ensure that the flow of information proceeds smoothly and uninterrupted between joint commands, U.S. commands, and NATO partner nations.

Coalitions continue to experience difficulties in sharing intelligence due to countries' differing policies on sharing of classified information, as well as technical hurdles for sharing information. The sharing of timely and accurate intelligence has increased with the use of the Battlefield Information Collection and Exploitation Systems (BICES), but BICES has not yet been widely adopted. Communication systems

---

[92] Lynn E. Davis and Jeremy Shapiro, eds., *The U.S. Army and the New National Security Strategy*, Santa Monica, Calif.: RAND Corporation, MR-1657-A, 2003.

[93] Nora Bensahel, "Preparing for Coalition Operations," in Lynn Davis, *The U.S. Army and the New National Security Strategy*, NATO "Senior Civilian Representative Report, A Comprehensive Approach Lessons Learned in Afghanistan," NATO, July 15, 2010. ISAF, "After Action Review Report: NATO-Afghanistan Transformation Task Force (NATTF)," HQ ISAF, Kabul, 2013, pp. 7–9.

and the interoperability between nations are still an issue, with secure communications continuing to be problematic. Additional recommendations made in these reports include providing for substantial and ongoing coalition participation in war games and more extensive planning and training, providing for coalition support requirements, and developing a database of coalition-ready forces.[94]

In addition to participating in coalition operations in Afghanistan, Iraq, and elsewhere, some U.S. allies, including the UK and Australia, have been active participants in developing concepts, doctrine, and studies that have yielded relevant insights for U.S. forces and future coalition operations concerning the operational environment, landpower, and economies of scale that can be achieved by coalitions. The Australian Army has published updated doctrine on landpower this year, and the UK Development Concepts and Doctrine Center (DCDC) has been another close collaborator in U.S. and allied efforts to formulate and assimilate the lessons of the past 13 years into doctrine and future projections, including through the Allied Command Transformation.[95] Australian Army Strategy chief brigadier Mick Ryan, who deployed to Afghanistan and was assigned to the U.S. Joint Staff, concluded that one of the key changes he witnessed was "the dawning appreciation of the capacity to influence populations as a core competency of military forces and senior leaders."[96]

NATO's ability to form and lead coalition command structures has also increased through the past 13 years. Britain has led the NATO ISAF command in Afghanistan; NATO led the Libyan Odyssey Dawn operations, and France took the lead in the Mali intervention. NATO's land component command has developed a deployable headquarters capability. Another little-known development has been the grow-

---

[94] Nora Bensahel, in Spirtas et al., 2008; and Senior Civilian Representative Report, pp. 13–14.

[95] The Australian Army, *Land Warfare Doctrine 1: The Fundamentals of Land Power 2014*, Australian Army Headquarters, 2014; UK Ministry of Defence, *Future Land Operating Concept: Joint Concept Note 2/12*, The Development, Concepts and Doctrine Centre, 2012.

[96] Brigadier Mick Ryan, "After Afghanistan: A Small Army and the Strategic Employment of Land Power," *Security Challenges*, forthcoming.

ing experience of U.S. SOF operating with coalition SOF. U.S. SOF have devoted significant effort over the past 13 years to developing the partner SOF capabilities around the world and operating increasingly with other countries' SOF. Twenty-four countries deployed special operations units to Afghanistan under the command of a rotating UK-Australian brigadier. The most extensive effort to institutionalize a coalition SOF capability is the formation of the NATO SOF Headquarters (NSHQ) in Mons, Belgium, which conducts year-round education and training courses that ensure interoperability and develop common standards and approaches to operations. After a series of exercises in 2014, NATO validated NSHQ's ability to deploy as a command headquarters. This capability could be employed for three to six months while NATO members or other countries prepare to deploy a longer-term command headquarters. NSHQ maintains a daily battle rhythm to provide the Supreme Allied Commander Europe with a ready and informed headquarters staff.[97]

Amid numerous rancorous debates about the lessons of the past 13 years, there appears to be wide agreement on the desirability and utility of multinational coalitions. That in turn would imply that investing in improved coalition effectiveness is worthwhile. As Lincoln Bloomfield has written, "even in the case of another sui generis, one-off contingency, there will be no need, and frankly, no excuse to treat it [a coalition operation] as a pick-up game when it comes to enlisting willing allies in the fight."[98]

---

[97] The information in this paragraph is from a RAND visit to NSHQ in May 2014 to receive briefings and observe training and education courses at Mons and Chievres, Belgium, facilities.

[98] Lincoln P. Bloomfield, Jr., "Brave New World War," *Campaigning: Journal of the Department of Operational Art and Campaigning*, Joint Advanced Warfighting School, Summer 2006, p. 15.

# CHAPTER FOUR

# Future Conflict and Implications for the JIIM

[A]ll the services regarded the counterinsurgency wars in Iraq and Afghanistan as unwelcome military aberrations, the kind of conflict we would never fight again—just the way they felt after Vietnam. The services all wanted to get back to training and equipping our forces for the kinds of conflict in the future they had always planned for: for the Army, conventional force-on-force conflicts against nation-states with large ground formations; for the Marine Corps, a light, mobile force operating from ships and focused on amphibious operations; for the Navy, conventional maritime operations on the high seas centered on aircraft carriers; for the Air Force, high-tech air-to-air combat and strategic bombing against major nation-states.

I agreed with the need to be prepared for those kinds of conflicts. But I was convinced that they were far less likely to occur than messy, smaller, unconventional military endeavors. . . . The war in Afghanistan, from its beginning in 2001, was not a conventional conflict, and the second war against Iraq began with a fast-moving conventional offensive that soon deteriorated into a stability, reconstruction, and counterinsurgency campaign—the dreaded "nation-building" that the Bush administration took office swearing to avoid. . . . Developing this broad range of capabilities meant taking some time and resources away from preparations for the high-end future missions the military services preferred.[1]

—Former Defense Secretary Gates

---

[1]    Gates, 2011, pp. 118–119.

This concluding chapter builds on the previous chapters' analysis to assess how the recent experience may be applicable to future conflict and what that might imply for the U.S. government, military, and potential partners. The first section reviews future trend assessments and finds that those assessments suggest that the lessons of the past 13 years retain their relevance going forward. The next section argues that this "new normal" of a continuing high incidence of irregular and hybrid warfare, whether conducted by states or nonstate actors, indicates the need for a theory of success that can serve as a compass for strategy in these conditions, where victory may be elusive but security solutions remain imperative. The third section recommends exploration of seven deeper institutional reforms that may better prepare the nation to confront threats successfully.

The notion that a more fundamental adaptation of the U.S. national security system is needed for this era is not new. The Joint Staff's *Decade of War* study posited that "the Cold War model that had guided foreign policy for the previous 50 years no longer fit the emerging global environment."[2] This argument holds that the U.S. approach to national security and the basic orientation of its military are rooted in an era characterized by state-on-state conflict among standing forces of nation-states. The historical survey of Chapter Two suggested that even before the Cold War, irregular forms of warfare were bedeviling the United States, which periodically experimented with new approaches to counter them. If true, this broad trend would suggest that the need for adaptation by the U.S. government and its military continues even as it draws down in Afghanistan. New conflicts, a Eurasian crisis, and a metastasized terrorism problem all loom, promising little respite from threats that may demand some type of response.

The lessons derived from the workshop and other research suggest that the policy process is not optimized to produce clear ends, efficacious ways, or adequate means and is inhibited by inadequate civil-military interaction at the levels of policy, strategy, and implementa-

---

2   JCOA, 2012, p. 1.

tion.[3] While adaptation certainly did occur, as noted in the previous chapter, it was slow, incomplete, and mostly ad hoc. Hew Strachan observed a certain tendency to overstate the adaptation, noting that "If the U.S. army had taken as long to change in the Second World War, the war would have been almost over by the time it had completed the process."[4] One important form of institutional change has occurred, however, in military doctrine, with several revisions of joint, Army, and Marine doctrine on counterinsurgency, stability operations, and special operations. The Marine Corps is producing a new version of its 1940 Small Wars Manual. Despite this, as the Gates observation above suggests, there is a certain bureaucratic tendency to view recent experience as an aberration rather than a chapter from which future adversaries will learn.

## Future Conflict Trends

A number of projections discussed in this section foresee an ongoing incidence of irregular and hybrid war. The risks of conventional and nuclear war nonetheless remain, and the rise of peer competitors may bring them to the fore in the more distant future. A new RAND study assesses that China's current level of integration into the global system means it is less likely to behave in an unconstrained manner and pose an existential threat as the USSR did in the Cold War. "China is not the Soviet Union, which created a separate and distinct sphere from the West, dominated it politically, and controlled it economically," the report states, and its conclusion "rejects the perspective that China

---

[3]    In addition to the workshop findings cited in Chapter Three, and Davidson (2013), the Gates memoir which chronicles his cabinet level view of two administrations between 2006 and 2013 enumerates in great detail the frequent and recurring frictions between the White House and military leaders, despite his efforts to offer compromises and serve as a bridge between the White House preference for minimalist approaches to its foreign policy objectives and the military's more expansive view of the measures required to achieve success.

[4]    Strachan, 2013, p. 241.

should be treated as a 21st-century Soviet Union."[5] Regardless of the likelihood of major conventional war with any of a number of possible adversaries, the magnitude of their consequences requires the joint force to retain the capabilities to conduct conventional war and deter a nuclear war.

The first overall trend is a decline in the number of conflicts, as noted by the National Intelligence Council's *Global Trends 2030: Alternative Futures* report and documented in several academic studies.[6] Among the conflicts that are occurring, *Global Trends 2030* and other reports identify a changing character of war in these conflicts, which is a tendency for those conflicts to be irregular and hybrid in nature. Two possible reasons for both state and nonstate actors to resort to irregular warfare are (1) the U.S. overmatch in conventional and nuclear capability and (2) because state and nonstate actors can often achieve their ends using these lower-cost means. In *Global Trends 2030*, the National Intelligence Council identified two changes in the character of conflict: First, it projected that "most intrastate conflict will be characterized by irregular warfare—terrorism, subversion, sabotage, insurgency, and criminal activities."[7] Intrastate conflict will also be increasingly irregular, noting that "[d]istinctions between regular and irregular forms of warfare may fade as some state-based militaries adopt irregular tactics." The latter was recently illustrated by Russia's actions in Ukraine where it employed proxies, non-uniformed personnel, and a variety of subversive, economic blackmail and cyber

---

[5]   Terrence K. Kelly, James Dobbins, David A. Shlapak, David C. Gompert, Eric Heginbotham, Peter Chalk, and Lloyd Thrall, *The U.S. Army in Asia, 2030–2040*, Santa Monica, Calif.: RAND Corporation, RR-474-A, 2014, pp. iii, 40.

[6]   Once such study, whose findings are described below, is Thomas S. Szayna, Angela O'Mahony, Jennifer Kavanagh, Stephen Watts, Bryan A. Frederick, Tova C. Norlen, and Phoenix Voorhies, *Conflict Trends and Conflict Drivers An Empirical Assessment of Historical Conflict Patterns and Future Conflict Projections*, unpublished RAND Corporation research, 2013.

[7]   National Intelligence Council, *Global Trends 2030: Alternative Worlds*, NIC 2012-001, December 2012, pp. 59–60.

deception techniques to annex the Crimea and foster uprisings without waging a conventional assault to achieve its ends.[8]

The trends also show that despite an overall decline in the number of conflicts worldwide, the United States has been increasingly involved in interventions since 1990. In addition, the majority of those conflicts in which it has been involved since 2000 can be characterized as "wars among the people." A RAND Arroyo Center study documented that U.S. forces have been engaged in intrastate conflicts against substate and nonstate actors. It concluded: "Based on our projections, the Army has to be ready for interstate conflict and needs to have the type of forces associated with fighting state actors, *but Army forces are more likely to be engaged in intrastate conflicts and its forces have to be ready for the operational environments typically associated with inter-group (ethnic, sectarian) conflicts and insurgencies.*"[9] These trends directly suggest that the experience and lessons of the past 13 years will remain highly relevant.

Another projection is that the diffusion of technology will produce more hybrid warfare, as less capable adversaries (state and nonstate) gain access to and the ability to use a variety of more potent weapons. Hybrid warfare has been defined as conflict with "an adversary that simultaneously and adaptively employs a fused mix of conventional weapons, irregular tactics, terrorism and criminal behavior in the battle space to obtain political objectives."[10] Thus actors will be able to wreak more damage with more powerful weaponry, compared to the relative low lethality of U.S. adversaries in the past decade. The Global Trends report says: "the spread of precision weaponry—such as standoff missiles—may make some conflicts more like traditional forms of warfare." The National Defense Panel Review of the 2014 Quadrennial Defense Review also identified "wider access to lethal and disruptive technologies" as a major trend and stated that "[d]iffusion of these

---

[8]    Janis Berzins, "Russia's New Generation Warfare in Ukraine: Implications for Latvian Defense Policy," *National Defence Academy of Latvia Policy Paper Number 02*, April 2014.

[9]    Szayna, O'Mahony, et al., 2013, p. 169.

[10]    Francis G. Hoffman, *Conflict in the 21st Century: The Rise of Hybrid Wars.* Arlington, Va.: Potomac Institute for Policy Studies, 2007.

technologies will enable regional states to put U.S. interests, allies and forces at risk, and will enable small groups and individuals to perpetrate large-scale violence and disruption."[11]

In his memoir, former Secretary of Defense Robert Gates also subscribed to this view of the future. He wrote: "By 2009, I had come to believe that the paradigms of both conventional and unconventional war weren't adequate anymore, as the most likely future conflicts would fall somewhere in between, with a wide range of scale and lethality. Militias and insurgents could get access to sophisticated weapons. Rapidly modernizing militaries, including China's, would employ 'asymmetric' methods to thwart America's traditional advantages in the air and at sea. Rogue nations like Iran or North Korea would likely use a combination of tactics."[12]

Table 4.1 depicts some of the more readily available technologies that have made their way into the hands of various nonstate actors. While the traditional method of diffusion has been state sponsorship, as in the classic example of Hezbollah supplied by Iran and Syria, state collapse, capture on the battlefield, theft, and black-market sales are other avenues by which lethal technologies may increasingly be obtained in the future. In addition, other nonlethal technologies are readily accessible and will provide a wide range of actors with the ability to increase their nonlethal effects substantially and thereby compound the impact of their lethal activities.

Diffusion of weapons of increasing lethality is just one trend that will empower adversaries of the future. For example, nonlethal technologies including unarmed aerial vehicles are readily available commercially, with micro unmanned aerial vehicles (micro UAVs) and swarming tactics expected to increase in the near future. Cyber tools and tactics also enable influence operations of increasing impact, with the added advantages of low barriers to entry and ability to operate anonymously. In the recent example of Russia's actions in the Ukraine and

---

[11]   William J. Perry and John P. Abizaid, *Ensuring a Strong U.S. Defense for the Future*, Washington D.C.: United States Institute of Peace, 2014, pp. 14–15.

[12]   Gates, 2014, p. 303.

elsewhere, it employed trolls, bots, and hacktivists.[13] Influence operations of global reach and near instantaneity greatly magnify the ability of adversaries to coerce, deceive, and subvert, as well as recruit, plan, and operate. Finally, criminal networks, their tactics, and proceeds are empowering groups and enabling them to operate as quasi-states.[14]

Diffusion of technologically advanced weapons—man-portable air defense systems (MANPADS), anti-tank guided missiles (ATGMs), and UAVs in particular—is occurring, although it is not clear if the rate of diffusion is increasing. It is estimated that 5,000–7,500 MANPADS are currently held by nonstate actors.[15] There are at least nine nonstate groups that possess first-generation ATGMs. These groups include Somali militiamen,[16] Hamas,[17] and Free Syrian Army rebels.[18] A smaller number of nonstate actors such as Hezbollah, for example, have recently obtained second generation ATGMs from Iran and Syria.[19] Following the collapse of the Gaddafi regime, an estimated 15,000 MANPADS were unaccounted for and suspected to have fallen into the hands of armed groups or terrorist units, like al Qaeda in the Maghreb, Hamas in Gaza, Boko Haram in Nigeria, or Syrian insur-

---

[13] "Trolls" are organized groups of people who leave messages or comments on websites for the direct purpose of shaping international opinion. Max Seddon, "Documents Show How Russia's Troll Army Hit America," *BuzzFeed World*, June 2, 2014.

[14] See, for example, Max G. Manwaring, *Gangs, Pseudo-Militaries, and Other Modern Mercenaries: New Dynamics in Uncomfortable Wars*, Norman, Okla.: University of Oklahoma, 2010.

[15] U.S. Department of State, "MANPADS: Combating the Threat to Global Aviation from Man-Portable Air Defense Systems," July 27, 2011; Eric G. Berman, Matt Schroeder, and Jonah Leff, "Man-Portable Air Defence Systems (MANPADS)," Research Note No. 1, *Small Arms Survey*, 2011, p. 3.

[16] United Nations Security Council, *Report of the Monitoring Group on Somalia Submitted in Accordance with Resolution 1853 (2008)*.

[17] Isabel Kershner, "Missile from Gaza Hits School Bus," *New York Times*, April 7, 2011.

[18] Charles Lister, "American Anti-Tank Weapons Appear in Syrian Rebel Hands," *Huffington Post*, April 9, 2014.

[19] Siemon Wezeman et al., "International Arms Transfers," *SIPRI Yearbook 2007: Armaments, Disarmament and International Security*, Oxford: Oxford University Press, 2007, p. 410.

gents.[20] The most significant recent example is of the Islamic State of Iraq and Syria (ISIS) capturing weapons and munitions that arguably have made it a "full-blown army."[21] Although the details of the weapons it possesses are unknown, a senior U.S. official states that ISIS possesses "advanced weapons from Syrian and Iraqi bases that they have overrun."[22] According to the secretary-general of the Kurdish Regional Government's Ministry of Peshmerga Affairs, ISIS took the weapons stores of four Iraqi army divisions.[23]

As dramatic as these recent examples are, they are anecdotal and alone do not establish whether the rate of diffusion is increasing. Some attempts to control the diffusion of technology have been at least partly effective, through both arms control and arms buyback programs such as the one initiated by the U.S. to recoup Stinger missiles given to Afghan mujahedeen in the anti-Soviet Afghan war of the 1980s.[24] Whatever the rate of diffusion, the consequences of diffusion are demonstrable. David Johnson describes the relative ease with which Hezbollah transitioned from less lethal forms of irregular warfare to hybrid warfare via a state sponsor that provided weapons and training.[25] In the Lebanon war of 2006, Israeli forces paid a very heavy price in casualties and confronted serious difficulties due to the fact that the small Hezbollah force was trained, organized, and armed with sophisticated weapons including ATGMs, middle and long-range rockets, and MANPADS—and

---

[20] Andrew Chuter, "5,000 Libyan MANPADS Secured: Some May Have Been Smuggled Out," Defense News, April 12, 2012.

[21] Testimony of State Department Deputy Assistant Secretary for Iraq and Iran Brett McGurk, Senate Foreign Relations Committee, July 23, 2014.

[22] Greg Miller, "ISIS Rapidly Accumulating Cash, Weapons, U.S. Intelligence Officials Say," The Washington Post, June 24, 2014.

[23] Nabih Bulos, Patrick J. McDonnell, and Raja Abdulrahim, "ISIS Weapons Windfall May Alter Balance in Iraq, Syria Conflicts," Los Angeles Times, June 29, 2014.

[24] Government Accountability Office, Further Improvements Needed in U.S. Efforts to Counter Threats from Man-Portable Air Defense Systems, May 13, 2004, p. 10.

[25] David E. Johnson, Military Capabilities for Hybrid War: Insights from the Israel Defense Forces in Lebanon and Gaza, Santa Monica, Calif.: RAND Corporation, OP-285-A, 2010; David E. Johnson, Hard Fighting: Israel in Lebanon and Gaza, Santa Monica, Calif.: RAND Corporation, MG-1085-A/AF, 2011.

because the Israeli Defense Force was not prepared to meet this threat. The United States enabled a similar transition in the 1980s when it supplied the Afghan mujahedeen with Stinger missiles, which had a decisive effect in escalating the costs to the Soviet forces occupying Afghanistan.

In summary, the continued high rate of irregular warfare among the conflicts that occur and those conflicts that the U.S. is involved in, plus the diffusion of lethal technologies to create higher incidence of hybrid warfare, create a continued demand signal for the United States to address both irregular and hybrid threats. (See Table 4.1.)

## The Need for a Theory of Success

This section posits that this "new normal" conflict environment of multiple, simultaneous challenges that differ substantially from those of the Cold War or World War II era requires a new way of operating that can achieve national security objectives at bearable cost. Successfully aligning ends, ways, and means, in turn, requires a theory of success suited to the current environment. This section outlines the argument for and characteristics of such a theory.

The need to retain and refine capabilities for irregular, hybrid, and conventional war as well as a nuclear deterrent runs up against the constrained resources of the present and future years. The U.S. joint force is undergoing significant downsizing due to the automatic budget-cutting legislation. This fact inevitably bears on the topic of this paper, which is the distillation and application of the policy and strategic lessons of the past 13 years to the future threats and challenges that the U.S. government and its forces may be called upon to address. Means affect the options available to the policymaker, and strategy must reflect that. Making strategy requires making choices. As the Quadrennial Defense Review (QDR) review panel and other commentators have noted, however, strategy should not be entirely budget driven; the former critiqued the single-war assumption on which the QDR's force-sizing construct is based. The QDR review panel does not argue for returning to the previous two-war construct but for taking into account a world of multiple threats that the U.S. military must be

**Table 4.1**
**Diffusion of MANPADS, ATGMs, and UAVs**

| Technology | Selected Models by Manufacturing Country | Selected Nonstate Diffusion |
|---|---|---|
| **MANPADS** | | |
| Passive infrared seekers | SA-7 Grail (Russia); SA-14 Gremlin (Russia); SA-16/118 (Russia); PIM-92 Stinger (U.S.); HN-5 (China) | Abkhazian Congregation of the Caucasus Emirate; al Qaeda cell (Kenya); al Qaeda in the Arabian Peninsula; al Qaeda in the Islamic Maghreb; al Shabaab; Burundi insurgents; Chadian Union of Forces for Democracy and Development; Chechen rebels; Democratic Republic of Congo insurgents; Revolutionary Armed Forces of Colombia; Hezbollah; Hizbul Mujahideen (Kashmir); Iraqi insurgents; Islamic Resistance Movement (Hamas); Islamic State of Iraq; Libyan Revolutionary Brigades; Lord's Resistance Army (Uganda); Mouvement national de libération de l'Azawad (Mali); Palestinian Islamic Jihad; Kurdistan Workers' Party (PKK) (Turkey); Popular Front for the Liberation of Palestine; Rassemblement des forces pour le changement (Chad); Shan State Army (Myanmar); Somaliland (unilaterally declared government); Sudanese Revolutionary Front; Syrian anti-government armed groups; Taliban (Afghanistan); United Wa State Army (Myanmar); Ukrainian rebels |
| Radio command line of sight | Blowpipe (United Kingdom) | Chechen rebels; Taliban (Afghanistan) |
| Laser-beam riding | SA-18 Grouse (Russia); RBS-70 (Sweden); Starstreak (U.K.); FIM-92C Stinger (U.S.) | None (only one report says Hezbollah and Hamas) |
| **ATGMs** | | |
| Manual command to line of sight; wire-guided | AT-3 Sagger (Russia) | Al Shabaab; Hezbollah; Iraqi insurgents; Libyan Revolutionary Brigades; Islamic Resistance Movement (Hamas); Syrian anti-government armed groups |

**Table 4.1—continued**

| Technology | Selected Models by Manufacturing Country | Selected Nonstate Diffusion |
|---|---|---|
| Semi-automatic command to line of sight; wire-guided; radio- or laser-beam riding | AT-4 Spiggot (Russia); AT-5 Spandrel (Russia); AT-7/AT-13 Saxhorn (Russia); AT-14 Spriggan (Russia); Missile d'infanterie léger antichar (MILAN) (France); BGM-71 TOW (U.S.) | Al Shabaab; Hezbollah; Iraqi insurgents; Libyan Revolutionary Brigades; Islamic Resistance Movement (Hamas); Syrian anti-government armed groups |
| Passive infrared, radar, or laser-guided | Type 01 LMAT (Japan); Javalin (U.S.); Spike (Israel); Nag (India); HJ-12 (China) | None |
| UAVs | | |
| Short range, low technology | Radio-controlled model airplanes | al Qaeda; Revolutionary Armed Forces of Colombia (FARC); Hamas; Hezbollah: Palestinian Terrorist Group |
| Short range, high technology | RQ-11 Raven (U.S.) | None. Low cost will be widely diffused in near future. |
| Long range, low technology | Ababil (Iran); Seeker 400 UAV (South Africa) | Hamas; Hezbollah |
| Long range (300 km), high technology | Wing Loong UAV (China); MQ-1 Predator (U.S.); MQ-9 Reaper (U.S.); RQ-7 Shadow (U.S.); X-47B (U.S.) | None. Widespread diffusion unlikely in near term. |

SOURCES: Data from International Institute for Strategic Studies, *The Military Balance 2014*, 2014; and Stockholm International Peace Research Institute (SIPRI), SIPRI Arms Transfers Database, 2014.

prepared to address. The report states: "Our concern is that the threats of armed conflict for which the United States must prepare are more varied than they were 20 years ago. In short, the logic behind the two-war standard is as powerful as ever, but we believe that logic should be expressed in a construct that recognizes that the U.S. military must have the capability and capacity to deter or stop aggression in multiple theaters—not just one—even when engaged in a large-scale war." [26]

The United States (and its allies) does not face an imminent, existential threat as it did during the Cold War, but that lack of a concentrating focus may make it harder to frame a national security strategy that correctly balances lower-order threats. Even if there is no existential threat, given the U.S. military's duty to safeguard the nation, the primary consideration for the joint force must be maintaining the ability to protect and defend the nation against dire threats. That requires a range of capabilities and a certain level of capacity, which are both being reduced through the automatic budget cutting mechanism that has been legislated. The pressure that this places on the force, and the Army in particular as the nation's largest service with the widest array of missions and capabilities, raises a very important question: If the currently planned cuts occur, must the Army's conventional forces reduce their focus almost entirely to being prepared to fight and win a major conventional war? If the answer is yes, then the burden of irregular warfare would shift to SOF (which includes the 50 percent of SOF that is part of the Army) and the other services. Yet SOF's force structure is not designed to meet the challenges of irregular warfare by itself, as all of its missions require combat support, combat service support, and particularly airpower and intelligence capabilities that do not reside within its ranks. SOF also represent a tiny portion of the joint force (33,000 uniformed operators) to confront what is considered the most likely form of current warfare. Moreover, the argument advanced above that a growing trend of the future may be hybrid warfare, would, if true, require capabilities of the entire joint force, including the Army.

The larger issue is whether the U.S. military as a whole accepts the thesis that the character of warfare is indeed changing, which would

---

[26]  Perry and Abizaid, 2014, pp. 24–26.

imply the need to change the way the U.S. military understands and approaches war. The changing character of warfare may be the meta-lesson of the past 13 years (or longer). If the character of warfare is changing, it poses a challenge to the joint force, and perhaps most profoundly the Army, as the nation's largest service with the widest range of capabilities, to continue to assess not only the threats of the future but also how it approaches them. This changing character of warfare could moreover be compounding a long-standing vulnerability in how the United States approaches war. Efforts to examine, clarify, and perhaps modify the approach to war would provide a sound basis for deciding what capabilities need to be preserved, refined, or created. War games and ongoing concept development are all grappling with elements of this issue, and a follow-on study to the *Decade of War, Vol. 1,* has been initiated.

The discussion at the RAND workshop and the ensuing analysis suggest that there is merit in examining and clarifying what would constitute a viable theory of victory that (1) fully accounts for the changed character of warfare, (2) reaffirms the ultimate political objectives of war, and (3) articulates what their achievement might look like. In other words, a theory of victory would serve as a compass for strategy.

The statistical trends outlined above identify irregular warfare as a major feature of the changing character of war, in addition to research carried out under a program by that name at Oxford University.[27] The case for expanding the view of war to include the political dimension is founded on classic war theory texts as well as a substantial body of analysis of the past 13 years' experience, including work cited in Chapter Three. Much of that work focused on the negative consequences of ignoring the political dimension of war. As summarized by Schadlow, "The root of Washington's failure to anticipate the political disorder in Iraq rests precisely in the characterization of these challenges as "post-war" problems, a characterization used by virtually all analysts inside and outside of government. The Iraq situation is only the most recent example of the reluctance of civilian and military leaders, as well as

---

[27] See Strachan (2013) for a description of the program and collected works, including two previous volumes.

most outside experts, to consider the establishment of political and economic order as a part of war itself."[28]

The QDR review panel also raised the political dimension in its
comments about multiple and proliferating forms of armed conflict
cited above. It did so as well in its description of challenges that ema-
nate from Russia and Iran and particularly their use of "political warfare." The report calls attention to "Iran's continued use of terrorism
and political warfare throughout the region" and "Russia's increasing
use of rapidly mobile and well-equipped special operations forces with
coordinated political warfare and cyberspace capabilities to create new
'facts on the ground,' particularly in areas of the former Soviet Union."[29]
This usage of the term *political warfare* refers to adversary tactics. The
term has also been used to describe a broader U.S. approach to national
security challenges.

This definition of political warfare was used by George Kennan:

> Political warfare is the logical application of Clausewitz's doctrine
> in times of peace. In the broadest definition, political warfare is
> the employment of all the means at a nation's command, short of
> war, to achieve its national objectives, to further its influence and
> authority and weaken those of its adversaries.[30]

Whether called political warfare or the political dimension of war,
accepting this broader definition of war would then require a theory
of victory to adequately account for that dimension. Political outcomes
would be embraced as a principle and articulated specifically in each
case. Winning battles does not ensure victory, and the elegant formula-
tions of Sun Tzu and Clausewitz are often quoted. But the translation
of this exhortation into the operative principle in U.S. strategy does not
routinely occur.

---

[28] Nadia Schadlow, "War and the Art of Governance," *Parameters*, Vol. 33, No. 3, 2003.

[29] Perry and Abizaid, 2014, p. 19.

[30] George Kennan, *Policy Planning Staff Memorandum 269*, Records of the National Security Council RG 273, NSC 10/2, Washington, D.C., May 4, 1948.

To summarize, the theory is based on a broader conception of war to include the political dimension. That broader conception in turn necessitates a theory of success that addresses the full dimensions of war. The ways of achieving success encompass a much wider range of actions. The desired political outcome may be obtained via containment or mitigation, formally negotiated settlements, informal power-sharing, or elections and constitutional charters that establish the basis for a new political order. These outcomes may be sought, of course, without waging war; but the point is that the goals of war must also encompass some such outcomes. Otherwise the United States runs the risk of winning battles but failing to achieve strategic (or even operational) success. This broader view does not make attaining success necessarily more difficult, but rather opens up a broader definition of what success may look like and a wider range of ways to attain it. The will of the enemy must be subdued but, as Sun Tzu noted, it may be possible to do so without fighting or in a number of unexpected ways as the adaptive adversaries of the United States try to do.

This theory of victory may be more appropriately termed a theory of success, given the military connotations of the word victory and the fact that success may be defined in more modest terms. The idea that there can be degrees of victory, or success, is not new. Colin Gray developed the idea of multiple sliding scales representing degrees of decisiveness and achievement.[31] Using this concept, J. Boone Bartholomees writes that "While the words [winning and victory] are often used interchangeably, they offer a unique opportunity to distinguish important gradations that exist in the condition of success in war. The assertion here is that victory will be essentially total and probably final; that it will resolve the underlying political issues. It is certainly possible, however, to succeed in a war without achieving everything one sought or resolving all the extant issues. Winning implies achieving success on the battlefield and in securing some political goals, but not, for whatever reason, reaching total political success (victory). Lesser

---

[31]  Colin S. Gray, *Defining and Achieving Decisive Victory*, Carlisle, Pa.: U.S. Army War College, Strategic Studies Institute, 2002.

levels of success reflect lesser degrees of battlefield achievement or lesser degrees of decisiveness in solving or resolving underlying issues."[32]

The current Army vision statement is framed around three words: Prevent, Shape, Win. They are expressed as different ideas, and Win connotes winning in war. But Prevent and Shape can also be ways to win.[33] If fully operationalized, they considerably expand the ways to win beyond the realm of combat. Such an approach will likely require nonmilitary means, so its implementation would require the joint force act in concert as a JIIM enterprise that is sufficiently adept in political warfare or "noncombat war" that this becomes an effective default option. The theory then serves as a forcing function to achieve the much discussed but partly realized unified government action.

This theory of success explicitly includes a preference for aiming to win in Phase 0 whenever possible. This has obvious appeal in that it avoids the cost of war and yet achieves security objectives. In one sense it is simply an obvious statement of rational preference. But incorporating this preference into intent in a formal theory of victory can have a transformative effect on the "American way of war." The U.S. military must remain prepared to fight and win major conventional war, retain a viable nuclear deterrent, but under this theory of victory a greater weight of effort and focus of attention moves to "prevent and shape" in the Army's Prevent-Shape-Win construct as a more efficient means of obtaining desired outcomes. In other words, it seeks to win by preventing and shaping, and makes the necessary investment and adaptation to accomplish that if at all possible. And as a second consequence, when those efforts fail and a major Phase III operation must ensue, it will be explicitly grounded in a political strategy that envisions the war-ending conditions that must prevail through some combination of military and non-military ways and means. In other words, "winning" must include some concept of consolidation to achieve lasting results.

---

[32]  J. Boone Bartholomees, "Theory of Victory," *Parameters*, Summer 2008, p. 28.

[33]  LTG Charles T. Cleveland, USASOC commanding general, shared this insight with the study team and used it in developing the idea of long-duration special operations "campaigns."

A theory of success may be difficult to define without reference to a specific threat or problem it seeks to address, but its elements could be described broadly as follows. A theory of success would aim in the first instance at prevention of all-out conflict through proactive means, and mitigation of conflict when all-out victory is infeasible or elusive. That is to say, such a theory must account for the continuum of politics and war and be profoundly realistic in setting objectives. This is not an argument for minimalist objectives that aim at purely expedient or temporary fixes; there must be an insistence on sustainable outcomes to warrant the expenditure of effort. General characteristics of "ways" or approaches that are able to achieve such effects are (1) persistent presence as a substitute for overwhelming force to win the contest of wills, (2) strategic patience to pursue a course that may only achieve effects over time, and (3) elements of inventiveness or surprise that use understanding of the political environment to perform a kind of jujitsu to gain advantage in lieu of overwhelming force. (In other words, the characteristics are in many respects the opposite of those Colin Gray used to describe the traditional American "way of war" cited in the previous chapter.) Ways or approaches with those attributes may achieve satisfactory ends with relatively modest investments, at least compared to the costs of major wars and investments in the armaments of strategic deterrence.

## Proposals for Institutional Reform

The rise of irregular threats and constraints on resources pose an acute dilemma for U.S. strategy, increasing the imperative to remedy the deficiencies of the past 13 years. More than ever, the United States requires new approaches that can achieve satisfactory outcomes to multiple, simultaneous conflicts at acceptable cost. It must become more agile in adapting its strategy as circumstances warrant, and it must improve its ability to work effectively with all manner of partners. The growing role of SOF represents a potential advantage of strategic import, but operational concepts and constructs must be further refined to supply a seamless array of options for the application of JIIM power.

Aspects of a new way of operating have already been under development, exemplified by an unprecedented use of SOF in conjunction with conventional forces as well as partially successful experiments in JIIM teaming to access a wider range of capabilities. Special operations and conventional forces combined in new ways to provide a suite of capabilities that neither possessed alone, and the combinations enabled operations at larger scale and in more conflicts than SOF could provide alone. The Army's new operating concept recognizes the need to continue to adapt and innovate to execute combined arms operations in a complex future operating environment. The concept for the first time recognizes special operations as a core competency and thus opens the way for greater experimentation in combined conventional-SOF operations and development and institutionalization of capabilities.[34] The concept does not limit this vision to one type of war or one phase of operations; it states that "interdependence, gained by the right mix of complementary conventional and special operations forces, enhances success throughout the ROMO [range of military operations] and all phases of joint operations." This opens up the possibility for future major innovations in flexible force combinations and ways to employ them.

Some of the recommendations below apply broadly to the national security strategy and the full spectrum of war, and others are more focused on habitual deficits, which is to say the capabilities for prosecuting irregular war and executing proactive and preventive approaches that obviate the need to conduct costly land wars. Whatever the capabilities, they will only be effective if harnessed to a viable strategy that is guided by a theory of victory, or perhaps more appropriately, a theory of success. This argument provides the basis for seven proposals for reform or retention of capabilities.

The seven areas addressed below were selected based on their criticality to addressing the changing character of warfare. The challenges identified in Chapter Three were mapped to the joint framework for

---

[34] RAND was provided with a draft version of the concept, published as U.S. Department of the Army, *TRADOC Pamphlet 525-3-1, The U.S. Army Operating Concept: Win in a Complex World*, Washington, D.C.: Government Printing Office, October 7, 2014.

capabilities: doctrine, organization, training, material, leader development and education, personnel, and facilities, as well as civilian interagency and intergovernmental and multinational capabilities.

In identifying both lessons and possible remedies through the
workshop, poll, interviews, and research, a "short list" approach was
taken to aid in prioritizing areas deemed essential to (1) successful execution of national security strategy in a basic way or (2) the prosecu-
tion of irregular and hybrid war. This limiting approach was applied
because the current challenge is how to improve the ability of a smaller
joint force and government to master irregular and hybrid warfare
while retaining the conventional and nuclear capabilities. This is not
just a U.S. military imperative; the U.S. government and its allies and
partners need to grapple with the changing character of warfare and
creatively apply resources in the most effective ways.

Technology is often viewed as the most likely way to produce
greater efficiency and effectiveness, but substantial investment is often
required to realize those gains. As Chapter Two noted in its survey of
the evolution of warfare, airpower, and lethal technology reduced the
need for massive armies and logistical tails to support vast quantities of
"dumb" ordnance. Intelligence, surveillance, and reconnaissance (ISR)
and precision weapons greatly increased SOF "man-hunting" capability. Changes in organization and personnel, as well as operational
innovation, may be equally useful in overcoming the strictures of budgetary constraints and closing the gaps revealed in past performance.
The last big organizational change in the Army was Brigade Combat
Team (BCT) modularity.[35] The personnel system has remained largely
unchanged, in its fundamental approach to branches and specializa-
tions, since the advent of the all-volunteer force. The recommendations
below suggest areas of possible evolution for further study.

---

[35] The Army is engaged in an effort at organizational transformation and capability develop-
ment/preservation under the regionally aligned forces (RAF) concept. According to Army strategic guidance for 2013, the RAFs "must maintain proficiency in wartime fundamentals but also
possess a regional focus that includes an understanding of the languages, cultures, geographies,
and militaries of the countries where they are likely to be employed. In addition, as part of their
focus on training, RAFs must be able to impart military knowledge and skills to others." AWG
report, *Army Strategic Planning Guidance 2013*, 2013, p. 6.

### Enhancing Strategic Competence

The lessons from the past 13 years of war formulated here as well as other studies and analyses suggest that steps are needed to enhance U.S. strategic competence.[36] Frank Hoffman, in an essay of that title, argues that there is not a single solution but three avenues for addressing the deficit. He writes: "There is much discussion these days about fixing America's strategic thinking deficiencies. . . . Contrary to some perspectives, process and structure is important in the development and vetting of both good strategy and policy. The solution set will require three inter-related components, Structure, Process, and Education."[37]

For the structural part of the solution, many options have been developed for improving the making of strategy, and many of them point to the National Security Council as the logical entity where these improvements need to take place. The Eisenhower era has been identified as a model with two possible variants. A National Planning Board could be re-instituted, composed of statutory members from selected Departments of the Federal Government, and could reside within the Executive Office of the President. Recreating this body and establishing its secretariat as a function of the National Security Advisor is one way to reintroduce the NSC staff to long-range and conceptual thinking.[38] An alternative to an interagency planning process run by the NSC staff is a dedicated NSC Strategic Planning Directorate. [39] This option could include not only staff from the various departments but also academic or policy experts.

The goal of reforming the NSC is to provide a locus for strategic policy development—studying problems, crafting options, and

---

[36]  Andrew F. Krepinevich, Jr., and Barry D. Watts, *Regaining Strategic Competence,* Washington, D.C.: Center for Strategic and Budgetary Assessments, 2009.

[37]  Francis G. Hoffman, "Enhancing American Strategic Competency," in Alan Cromartie, ed., *Liberal Wars,* London: Routledge, forthcoming, pp. 6, 18.

[38]  Paul D. Miller, "Organizing the National Security Council: I Like Ike's," *Presidential Studies Quarterly,* Vol. 43, No. 1, September 2013.

[39]  Peter Feaver and William Imboden, "A Strategic Planning Cell on National Security at the White House," in Daniel W. Drezner, ed., *Avoiding Trivia: The Role of Strategic Planning in American Foreign Policy,* Washington, D.C.: Brookings Institution, 2009.

exploring alternatives. Regardless of the specific form that NSC reform takes, the NSC staff should remain limited to policy development and interagency coordination. These proposals do not override the cabinet and constitutional lines of authority: They do not entail the NSC staff assuming a role in policy execution or operational control over implementation.

In terms of education, it is important to remedy the deficit of strategic education on the civilian side, and on the military side to develop a more realistic and less rigid concept of how policy is made as part of the military education and planning. The National War College has introduced courses that could be a basis for providing enhanced strategic education to both civilian policymakers and staff and the senior military planners and strategists. Practices that the U.S. military has developed over the past 13 years can be employed at the policy level for developing an integrated political-military national strategy, as well as for contingencies in particular countries or regions. For example, the problem-framing and iterative elements of design can also be introduced into the policy planning process as a matter of course, to enforce rigor in identifying and revisiting assumptions, risks, and costs. Other measures could include routine use of a deliberate process for testing the range of policy options through interagency gaming and tabletop exercises.

## Organizational Adaptation

Transformations in organization and personnel may be especially critical to remedy the deficits revealed by past experience and prepare to meet the challenges of the future. Many innovations of the past decade can help point the way forward. Flat, adaptive, networked modes of operating have become a new organizational model for nonstate actors, businesses, and other entities, driven in part by technology that permits rapid acquisition and diffusion of information. The joint force may find it advantageous to examine options for radical change in its organization and personnel systems. Task organizing and mission command are two concepts that can drive further organizational adaptation.

Task organizing is a model for adapting structures to specific purposes, and exploration of deeper modes of task organization might yield greater flexibility and adaptability. All services have formed task forces, and joint task forces are the norm in deployed formations. The formation of task forces just prior to deployment contains certain limitations in that units are not routinely trained or prepared to work with those specific partners in the required manner. A model that incorporates task organization as a fundamental principle and implements it at lower echelons might foster more inherent traits of adaptability and flexibility.

Historically, the Army has changed its organizational structure based on the threat environment, from a divisional/corps structure built to counter the Soviet Union to Combined Arms Battalions (such as the 173rd Battalion (BN) Task Force (TF), built for operations in the Balkans). Modularity has been critical to Army successes, depicted through the number of TF-sized elements functioning at levels smaller than the BCT throughout operations in Iraq and Afghanistan. The doctrine for the BCTs—*FM 3-90.6 Brigade Combat Team*—states that the BCTs are still the "Army's combat power building blocks for maneuver, and the smallest combined arms units that can be committed independently."[40] In practice, however, this is not necessarily true. Combat operations over the past 13 years have demonstratively shown that U.S. Army forces routinely deploy and, more specifically, operate at levels below the BCT.

Guidance from Army senior leaders and multiple strategic guidance documents address the need to organize the force appropriately, stating that Army forces must be tailorable and scalable to specific missions and they must be prepared to respond rapidly to any global contingency mission.[41] One way that the Army is operationalizing the intent for tailorable and scalable forces is epitomized in the Regionally Aligned Force (RAF) concept. This concept is a total-force undertaking in which designated units must maintain proficiency in warfighting fundamentals but also take on additional training with a regional focus to include language, culture, geography, and regional militaries where they are to be employed.[42] These forces provide deployable

---

[40]  U.S. Department of the Army, 2006.

[41]  U.S. Department of the Army, *Army Posture Statement 2014*, 2014, p. 7.

[42]  U.S. Department of the Army, *Army Strategic Planning Guidance 2013*, 2013, p. 6.

and scalable regionally focused Army forces that are task organized for direct support of geographic and functional combatant commands and Joint requirements.

The 2nd Brigade, 1st Infantry Division was the first unit designated as the RAF and began actively supporting AFRICOM in March 2013.[43] The brigade deployed a wide range of force packages— ranging from small teams of soldiers, up to a BN TF-sized element— across a gamut of mission types in support of AFRICOM requirements. The Army is still examining the experience of these initial deployments and refining the RAF model. The service-retained RAF model requires a lead time of up to six months to supply the unit, whereas the allocated RAF model provides the unit for one year and allows it to collocate with the theater command for regional orientation and training. The unit provided is the BCT so the needed enablers must be sought through a request for forces. Expertise on foreign weapons systems is also limited, since only Special Forces are trained in their use.[44]

The U.S. Marine Corps and the special operations community routinely task-organize to battalion level and below. U.S. SOF deploy most often in very small formations, and in this regard the Iraq and Afghanistan experiences with 5,000 or more deployed at once in the same country are anomalous. Nonetheless, even though units were highly distributed and operated with a high degree of autonomy from their headquarters. Their task organization is also highly joint. SOF is organized in service component commands under SOCOM, which was created by congressional act after special operations units were unsuccessful due to inadequate jointness in the attempted rescue of the American hostages in Iran in 1980 (and a subsequent poor performance in Operation Urgent Fury in Grenada in 1983). Since then the joint ethos of the special operations community has become fairly well developed at the tactical level where joint air-ground teams are the norm. Operational commands have been heavily joint as compared to the largely service-specific command structures, such as Army bri-

---

[43] U.S. Department of the Army, *Army Posture Statement 2014*, 2014, p. 9.

[44] This information comes from RAND interviews conducted with U.S. military personnel at the U.S. Army Combined Arms Command, Fort Leavenworth, Kan., August 2013.

gades, divisions, and corps, which deploy in those formations with joint personnel augmentees.

The United States Marine Corps (USMC) Marine Air-Ground Task Force (MAGTF)—the principal organization upon which this force is built—is a highly scalable organizational model composed of command, ground, aviation, and logistics elements that vary in size and capability according to their assigned or likely missions and are specifically equipped for deployment by sea or air.[45] The smallest ground element, a Marine Expeditionary Unit, is a battalion-sized force that is combined with aviation, logistics enablers, and maritime assets if needed to perform the required mission. Larger ground elements are Marine Expeditionary Brigades and Marine Expeditionary Forces.

Historically the Marine niche has been expeditionary amphibious operations, but the Marines deployed frequently in large numbers throughout the Iraq and Afghan wars. There is a move now back to smaller-scale expeditionary roles. The Special Purpose Marine Air-Ground Task Force for Crisis Response (SP-MAGTF-CR) based at Moron Air Base in Spain was established in response to the 2012 attack on the American diplomatic compound in Benghazi, Libya, and the recent turmoil that has arisen in Mali, Algeria, and other North African countries.[46] Designed to support U.S. and partner security interests throughout the CENTCOM and AFRICOM theaters of operation, primary missions include embassy reinforcement, noncombatant evacuation operations, humanitarian assistance/disaster relief, and tactical recovery of aircraft and personnel.[47]

Originally comprised of 500 marines and sailors capable of responding to a myriad of crises, this unit differs from other SP-MAGTFs in that its organic aviation capability (MC-22 Ospreys and KC-130J) allows it to move large distances. The unit has conducted training missions with the French Foreign Legion, Spanish paratroopers, and Italian forces. Recent activity saw the unit deploy to East Africa

---

[45]   U.S. Department of the Navy, *Marine Corps Operations, MCDP 1-0*, August 9, 2011, pp. 2–6.

[46]   "3-Star Details New Marine Crisis-Response Force," *Marine Corps Times*, April 21, 2014.

[47]   General James F. Amos, *2014 Report to Congress on the Posture of the United States Marine Corps*, House Armed Services Committee, March 2014.

in support of an embassy evacuation mission that had the unit work in combination with the East African Response Force (EARF), Navy Seals, and U.S. Air Force assets. A sub-component of SP-MAGTF-CR, SP-MAGTF-Africa 13 is forward-based at Naval Air Station Sigonella, Italy, and consists of a company-sized Marine ground combat element that engages with partner militaries throughout Africa.[48] The unit has focused on training Ugandan and Burundian militaries engaged in operations to counter violent extremist groups including al Qaeda affiliates throughout the Maghreb region and to support the African Union Mission in Somalia.

The joint force might explore additional approaches to task organization. As the largest force, the Army faces the widest array of demands. Although the Army must retain the ability to conduct and prevail in major combat operations, it faces a competing demand that it perform equally well in other missions that are likely to constitute its most frequent form of employment. Despite the introduction of the RAF, the Army is still fundamentally based on the BCT, and it has adapted the BCT to form Advise and Assist Brigades (as it did in the latter years in Afghanistan) and battalion or company size advisory teams (less successfully, judging from the Iraq experience of military transition teams and other ad hoc advisory teams). An alternative model might be a multifunctional brigade (still qualified in decisive action) that readily produces a multifunctional battalion or company. Whether applied at the brigade or battalion level, this model would also eschew the BCT and enabler distinction in favor of a mission-based task organizing principle that would present a package that includes the "enablers" and even places them in the lead when the mission dictates a multifunctional unit with civil affairs, engineering or influence operators (MISO) as its main effort. Another option would be to form some advice and assist brigades (or smaller units) and assign them to geographic regions with high and sustained demand.

In the near term, the following recommendations could be used by the Army to create smaller, mission-ready units organized to accomplish a specific mission for a Combatant Commander instead of uti-

---

[48] Amos, 2014.

lizing the BCT model for deployments: (1) Limit the number of units that need to be trained and equipped to deploy under the BCT level. (2) Start the training earlier and ensure NTC or Joint Readiness Training Center rotations for the battalion-level joint task force prior to a deployment. Ensure that all enablers are present and the task force is assembled before the training rotation begins. (3) Obtain dedicated air assets for mobility and close air support for echelons below the BCT level. (4) Create the habitual relationship with the units that support each other. This requires early identification of air elements to provide sufficient time for adequate predeployment training.

## SOF-Conventional Force Interdependence

The recent robust use of special operations suggests the possibility of a new model, or models, for achieving operational or even strategic effect through a campaign approach. This represents a potentially potent new form of landpower that, if applied with strategic patience, can address threats without resort to large-scale military interventions. SOF-led campaigns can provide low-visibility, high-return security solutions in numerous circumstances. SOF have begun to develop the operational level art, planning, and command capabilities to realize this potential, but several additional steps are needed.

As noted above, the significant increase in reliance on SOF over the past 13 years has been recognized in the new Army operating concept as constituting a core competence for the service. A concomitant recognition in a joint concept would pave the way for development of the doctrinal, organizational, and other capability implications of this trend. The promise it holds is for a strategic approach that, in many future areas of operation, achieves lasting effect with fewer troops and other resources, though applied in some cases in a sustained and geographically distributed manner. To achieve its intended effect, such "persistent presence" must be appropriately conceived, organized, and led.

In the operating concept, the Army envisions three types of operations: SOF-centric, SOF-conventional-force blends, and more traditional operations led and conducted by a preponderance of conventional forces.[49] To lead the first two types of sustained operations, an argument can be made for further innovation in SOF command structures as well as hybrid SOF-conventional commands. Given that these environments and operations will occur more often than large scale conventional conflict, it suggests that these two types of operations may in fact be important for addressing most conflicts of the future. The need for deployable operational level SOF commands became apparent over the past 13 years as SOF became increasingly capable of leading sustained, country-wide or regional operations, in contrast to their previous employment in primarily tactical and short-term modes. To conduct these types of operations, SOF have relied heavily on ad hoc commands substantially filled by individual augmentees through a joint manning document.

SOF experience in the past 13 years suggests the need for more standing joint deployable command capability for SOF.[50] In recent years, SOF experimented with creating ad hoc operational level commands at both the one-star and two-star levels. After creating a two-star ad hoc unified SOF command for Afghanistan, the Special Operations Joint Task Force–Afghanistan (SOJTF-A) in 2013, SOCOM conducted an after action review and confirmed the utility of such a command to orchestrate all SOF deployed in a given theater of operations. The review also noted that effectiveness of future such commands could be improved by creating either a standing core of a deployable command or a full joint command well in advance of deployment to provide sufficient time to overcome the inherent drawbacks of an ad hoc

---

[49] The U.S. Army Training and Doctrine Command has just developed the next iteration of the Army operating concept. The RAND research team reviewed a draft version of the Army operating concept that was published as U.S. Department of the Army, *TRADOC Pamphlet 525-3-1, The U.S. Army Operating Concept: Win in a Complex World*, Washington, D.C.: Government Printing Office, October 7, 2014.

[50] The special operations counterterrorism joint task force is currently the only standing joint deployable special operations command capable of commanding operational-level missions, and it too is augmented through a Joint Manning Document. For other SOF, until the formation of a one-star command in the final year of Iraq operations, countrywide SOF operations in both Iraq and Afghanistan were commanded by an O-6 level command based on Army Special Forces Groups between 2001 and 2010.

staff.[51] Similarly, USASOC's Silent Quest exercise in 2013 concluded that a standing deployable SOF command structure was desirable.[52] To address the deficit, USASOC is in the process of creating a two-star 1st Special Forces Command (Airborne) (Provisional) to provide a deployable headquarters that integrates special forces, MISO, civil affairs and sustainment units[53] In addition, in 2013 the theater special operations commands were reassigned from the geographic combatant commands to SOCOM with the rationale that they would be provided increased special operations manning and resources to serve their doctrinal functions of planning and conducting special operations in a given theater.[54]

During the past 13 years, special operations and conventional forces made significant strides in operational coordination, beginning with a steep learning curve to share information and deconflict operations. SOF also achieved notable successes in precision targeting of high-value individuals due to the creation of interagency fusion cells that included all-source intelligence analysts and a vast array of technical intelligence means. Those methods have migrated in good part to the conventional forces, which employ the find, fix, finish, exploit, analyze, and disseminate methodology and which have integrated air ground surveillance and targeting at the tactical level, as demonstrated in the 2008 battle of Sadr City.

SOF have demonstrated the ability to conduct sustained operations in multiple theaters over the past 13 years, but their size limits them to deploying no more than some 10,000 operators at any one

---

[51] SOCOM Lessons Learned Operational and Strategic Studies Branch, "Special Operations Joint Task Force-Afghanistan (SOJTF-A): From Concept to Execution . . . The First Year," 2013. RAND has also conducted two classified assessments of the SOJTF-A.

[52] USASOC, "USASOC Silent Quest Facilitated War Game 14-1 Executive Summary and Final Report," 2014. It also states that "mission command constructs in the future will have to integrate elements of SOF, CF, IA and other partners, and be more flexible in order to address emerging requirements or new conflicts."

[53] If authorized by the U.S. Army, this provisional command would transition from the U.S. Special Forces Command to a deployable Modified Table of Organization and Equipment command.

[54] McRaven testimony, 2014.

time worldwide. The historic growth of SOF is now leveling off as budget cuts take their toll. Their limited size and the continued high demand signal warrant greater effort to create more flexible combinations of special operations and conventional forces to permit larger-scale operations and, perhaps most often, greater numbers of sustained, distributed operations. This would suggest that the next step in the evolution of the SOF-conventional force relationship could be a hybrid SOF-conventional command. The U.S. Army Special Operations Command's vision document, *ARSOF [Army Special Operations Forces] 2022*, proposed exploring this option, and its Silent Quest wargame concluded that "SOF and its partners must establish hybrid structures that include elements of SOF, CF and JIIM partners, and institutionalize these structures as part of its steady state organizational framework."[55] Over the past decade Army Special Operations Forces brigadier generals have served as assistant commanding generals in conventional divisions, a practice which provides senior SOF leaders with experience in conventional commands and opportunities to develop wider relationships outside the special operations community. The conventional force commanders, for their part, learn more about special operations perspectives and approaches. A hybrid command would provide an avenue for moving from intra-SOF unity of command to full joint unity of command at the operational level.

Regular and habitual teaming with conventional forces and joint SOF, perhaps through RAF, will increase and sustain such familiarity at the unit and individual level. Without such continued collaboration the past 13 years' experience of deconfliction, fusion cells, combined operations and other interdependent operations will soon become a distant memory. Numerous structures and practices can foster continued and increased SOF-conventional force interdependence across the range of military operations. One option for deepening and sustaining SOF-CF interdependence would be to revive the Military Adviser Training Academy (MATA) created during the Vietnam War and based at the Special Warfare Center and School at Fort Bragg, North

---

[55] U.S. Army Special Operations Command, "ARSOF 2022," 2013, and "USASOC Silent Quest Facilitated War Game 14-1 Executive Summary and Final Report," 2014.

Carolina. The Special Warfare Center and School has the required training capability and personnel to develop and operate such a course. A school at that major Army base, which is located a short distance from major Marine Corps installations, could provide a location for retention of advisory knowledge, education courses and training facilities for a joint, interagency and ministerial full-spectrum advisory and stability operations complex. Corps, division and brigade conventional units are based there, as well as a multitude of special operations units including civil affairs and psychological operations units.

As mentioned above, SOF have developed new doctrine and revised existing doctrine to take account of the lessons learned and practices developed over the past 13 years.[56] A second edition of the USASOC Planner's handbook of Operational Art and Design was published in September 2014, with additional development of operational design and illustrative historical vignettes to aid SOF in planning long-duration SOF-centric campaigns. In addition, USASOC's Assessing Revolution and Insurgent Strategy Project has published four in-depth studies on insurgent, revolutionary, and resistance warfare.[57] Its annual Silent Quest cycle of games include a wide array of joint, interagency, and multinational participants. Finally, special operations personnel are now assigned to Fort Leavenworth to participate in all-corps and division-level exercises and supply additional courses and education in special operations to the School of Advanced Military Studies. To achieve the intended effect, these initiatives all need to be maintained, extended, and further developed.

**Innovative and Multifunctional Personnel**

Personnel requirements for the future are that individuals be innovative and multifunctional. It is not sufficient that doctrine enunciate that decentralization in the form of mission command or that education

---

[56] In addition to the revised version of Joint Publication 3-05, Special Operations, in July 2014, Army Doctrinal Pamphlet 3-05 was published in August 2012.

[57] USASOC, Assessing Revolutionary and Insurgent Strategies, *Casebook on Insurgency and Revolutionary Warfare: 23 Summary Accounts; Casebook on Insurgency and Revolutionary Warfare, Volume II 1962–2009; Human Factors Considerations of Underground in Insurgencies, 2d Edition*, 2013; and *Undergrounds in Insurgent, Revolutionary and Resistance Warfare, 2d Edition*, 2013.

and training inculcate the value and method of taking general guidance and devising the most effective means of accomplishing the commanders' intent. The personnel system itself must reward innovation.

Two steps have been taken to increase incentives and means to encourage innovation at the individual level. The new Army officer evaluation report (OER) includes "creativity" as a character trait to be discussed. This is a valuable first step but much more could be done to mitigate the inherent conformist tendencies of a hierarchical organization in which superiors control the career path of subordinates. An interesting development is the Army's adoption of a 360-degree general officer online evaluation form, in which any general officers are allowed to anonymously rate their peers. This rating tool is used strictly for developmental purposes for the officer to learn and adapt, rather than an assessment used for promotion evaluation.

Increasing multifunctional personnel may be another way to develop expertise within a shrinking force. While a shrinking force may not be able to accommodate large numbers of personnel trained in only one specialty, several models exist for multifunctional personnel with more than one specialty. The Foreign Area Officer specialty is one, and the Marine Corps Foreign Security Force Adviser is another. These trained advisory personnel could also serve as training cadre in a revived MATA advisory academy to produce larger numbers of advisers when needed. Finally, the joint force could adopt the Special Forces habit of routinely cross-training soldiers on teams in each other's specialties.

Plans for surging expertise in needed regions can also help produce needed capability rapidly. To overcome the deficit of expertise and create a sustained pool of available manpower for advisory work in Afghanistan and Pakistan, the AfPak Hands program was created by the Chairman of the Joint Chiefs of Staff. Members from all four military services could sign up for two deployed tours to the region. Prior to deploying they received language and cultural training in a degree-based education program, and in between deployments they were assigned to relevant positions in the United States addressing the Afghanistan-Pakistan policy and program objectives. This program could be studied as a potential model for a future surge advisory capacity. For such a program to attract the best talent, participants' careers must not be derailed by making the commitment. Ideally, such service should be incentivized through the promotion or preferred assignment systems.

### Joint Capabilities for Irregular Warfare

While defense guidance does state that expertise and capabilities will be maintained for conducting small-scale counterinsurgency and stability operations, the precise level required should be derived from the current military plans in order to drive the appropriate resourcing, force structure, training, education, and equipping decisions.[58] A RAND study found that the diverse competencies were required for U.S. Army officers to operate effectively in a JIIM environment, concluding that: "Successful performance in joint, interagency, or multinational contexts requires the application of highly developed functional expertise to novel situations" and that "the JIIM domains are qualitatively distinct, though overlapping."[59] These distinct capabilities are highly fungible across many IW missions at various scales, as well as in security cooperation missions. To accurately assess the demand signal, military plans' Phase 0 operations should be surveyed to ensure development and retention of adequate competencies and specialized force structure.

Many organizations and training programs aimed at creating the necessary skills for conducting stability operations and building partner capacity, as well as preparing personnel to work in deploying interagency coordination cells and teams have been disbanded as the U.S. war in Iraq ended and the force deployed to Afghanistan declined.[60]

---

[58]   Interview, Joint Staff J-7, July 2014.

[59]   See also M. Wade Markel, Henry A. Leonard, Charlotte Lynch, Christina Panis, Peter Schirmer, and Carra S. Sims, *Developing U.S. Army Officers' Capabilities for Joint, Interagency, Intergovernmental, and Multinational Environments*, Santa Monica, Calif.: RAND Corporation, MG-990-A, 2011.

[60]   For example, naval, air force, and marine advisory programs have been terminated. The marine advisory program has been folded into the Marine Corps Security Cooperation Group. The training programs for deploying civilians at the Foreign Service Institute and Camp Atterbury have been disbanded. The Army retains an advisory training unit, the 162nd Brigade at Fort Polk, Louisiana, manned by some 200 soldiers. The Marine Corps

These special skills cannot be created overnight, and simply maintaining a library of lessons learned and programs of instruction will not produce optimal results. For maximum effectiveness and readiness, it can be argued that actual capability must be retained at some scale across the services. A joint entity such as the Joint Center for International Security Force Assistance (JCISFA) may be the most efficient method of articulating the demand for and tracking the capability and lessons of stability and advisory missions, but some service-specific capability is needed, for example to train foreign maritime, naval, and air forces on platforms and techniques other than those used by U.S. forces. In a July 2014 report, the Government Accountability Office examined the process for tracking needed advisory specialties and recommended further improvements. Table 4.2 shows some of the relevant capabilities for irregular warfare and their current status. Funding for many of the remaining capabilities is not in the services' base budgets but rather in Overseas Contingency Operations funding, which is declining.

## Interagency and Intergovernmental Coordination

The shortage of civilian interagency capability and capacity was addressed in the previous chapter, as was the impediments to their movement at the tactical level in hostile environments, which in turn restricted their ability to complete their mission. The first of these constraints may not have a solution in the current budget environment, and after the killing of a U.S. ambassador in Benghazi the domestic political environment in the United States has made the State Department even more security conscious and risk averse.

Some studies recommend the creation of adequate civilian capability and capacity, which would require congressional support and funding for the State Department to revive the interagency CRC, adequately fund its parent organization, the Bureau of Conflict and Stabilization Operations, support the Civilian Stabilization Initiative, and consider creating a school to train civilians in reconstruction and stabi-

---

is doubling its civil affairs capacity by two battalions, but the Army is disestablishing one of the two civil affairs brigade commands (the 85th Brigade headquarters). Source: RAND interviews with joint staff, USMC, and USAID officials.

**Table 4.2**
**Selected Joint and Interagency Capabilities**

| Program | Status |
|---|---|
| Civilian Response Corps (CRC) | Likely Canceled |
| CRC predeployment training | Canceled |
| CRC interagency participation | Reduced |
| Marine Corps Security Cooperation Group (MCSCG) | Intact |
| USMC Advisory Training Group | Absorbed into MCSCG |
| Marine Corps Regional Culture and Language Familiarization Program (RCLF) | Intact |
| U.S. Air Force Air Advisor Academy AFA | Canceled |
| Maritime Civil Affairs and Security Training Command (MCAST) | Canceled |
| Human Terrain Teams | Under Review |
| Navy Expeditionary Combat Command (NECC) | Under Review |
| U.S. Army 162nd Brigade | Reduced |
| U.S. Army Asymmetric Warfare Group | Unknown |
| U.S. Army 85th Civil Affairs Brigade | Likely Canceled |
| Cultural Support and Female Engagement Teams | Unknown |
| U.S. Army IW Fusion Center (formerly COIN Center) | Canceled |
| Building Partnership Functional Capabilities Board | Canceled |
| Center for Complex Operations (NDU) | Reduced |

SOURCE: RAND interviews with Joint Staff J-7, July–August 2014, and additional research.

lization missions.[61] Several RAND studies have assessed the difficulties of providing large numbers of civilians with regional or other required expertise who are ready to deploy.[62]

---

[61]  See Dobbins, 2010; and Bensahel, Oliker, and Peterson, 2009.

[62]  Terrence K. Kelly, Ellen E. Tunstall, Thomas S. Szayna, and Deanna Weber Prine, *Stabilization and Reconstruction Staffing: Developing U.S. Civilian Personnel Capabilities*, Santa

Given these impediments, the requirement for civilians to be deployed at the tactical level might be more tightly scoped to address the issues identified. The most important contributions of civilian personnel may lie in their expertise, and as subject matter experts they may be able to provide the best service crafting political strategy and a unified approach employing all elements of national power. They may be best able to perform this strategic role as part of commands (such as U.S. Southern Command, which created a civilian "deputy to the commander" position) or as the foreign policy advisers currently assigned to most military commands, and in their traditional roles on country teams. They may also have greatest impact advising senior levels of host governments rather than employed in the field tactically to dispense aid, build clinics and schools, or providing services best performed by the local population themselves. To that end, the DoD Ministry of Defense Advisors (MODA) program provided senior DoD civilians to serve as advisers in Afghan ministries. If civilians are truly needed at the tactical level, they should be embedded in organizations that can provide for their safety and mobility, such as the PRTs and SOF teams. The purview of the Department of State diplomatic security service should be examined to determine whether current procedures unduly restrict these types of collaborative formations and assignments.

## Improving Coalitions and Leveraging Multinational Expertise

Policy and planning documents frequently assume that in the future the U.S. military will operate as part of a multinational coalition, as it did in Afghanistan with the NATO-led International Security Assistance Force (ISAF) and in Iraq with the Multinational Force–Iraq (MNF-I), a U.N.-sanctioned "coalition of the willing." If it is correct to assume that most U.S. deployments of significant scale and/or duration will occur as part of a coalition (Libya is the most recent example that suggests Iraq and Afghanistan are not outliers), an effort to

---

Monica, Calif.: RAND Corporation, MG-580-RC, 2008; and Thomas S. Szayna, Derek Eaton, James E. Barnett, Brooke Stearns Lawson, Terrence K. Kelly, and Zachary Haldeman, *Integrating Civilian Agencies in Stability Operations*, Santa Monica, Calif.: RAND Corporation, MG-801-A, 2009.

improve both coalition warfare and how the United States participates in coalitions could create more effective coalitions and thereby lessen the burden on U.S. forces. The coalitions of the past 13 years have been primarily based on U.S. personnel and structures, but coalitions of the future might rely more heavily on others, especially in those countries or regions where U.S. partners have a strong interest.[63] U.S. capabilities and resources will continue to make it a major contributor to many coalitions, especially as its NATO partners continue to struggle to meet the agreed level of defense spending (two percent of gross domestic product).

The United States can improve and emphasize its own coalition education, training, and operational practices. International attendance at U.S. professional military institutions remains significant, but funding for the State Department's International Military Education and Training (IMET) program has been reduced. U.S. professional military education curricula could increase its emphasis on coalition operations. The NATO Center for Lessons Learned has produced several studies to capture lessons from recent coalition experiences, which would be useful guides to needed changes as well as required reading for U.S. professional military education.[64] More robust pre-deployment training would also be desirable; often U.S. forces did not meet multinational units that they would serve alongside until a weeklong mission rehearsal exercise just prior to deploying.[65]

---

[63] This idea of expanded reliance on coalitions and international partners has been described as a Global Landpower Network, in LTG Charles T. Cleveland and LTC Stuart Farris, "Toward Strategic Landpower," *Army Magazine*, July 2013, pp. 20–23.

[64] NATO Senior Civilian Representative Report, A Comprehensive Approach Lessons Learned in Afghanistan, NATO 7/15/2010; *After Action Review Report: NATO-Afghanistan Transformation Task Force (NATTF)*, HQ ISAF, Kabul, 2013. These studies as well as the JCOA, *Decade of War*, Vol. 1, also note difficulties that arose when coalition partners were assigned to inappropriate tasks, unable to perform assigned duties due to information classification, or otherwise restricted in functions due to national government caveats.

[65] Principal author's personal observation at multiple predeployment training for one-, two-, and three-star commands deploying to Afghanistan in 2009–2011; some attended NATO training held for coalition members in Europe. On multinational training, see also Spirtas et al., 2008.

More international immersive experiences would also benefit U.S. personnel, relatively few of whom attend foreign military schools such as the NATO Defense College (NDC) in Rome. In such an environment, U.S. personnel are immersed in an international community of actual and potential coalition partners where U.S. personnel are in the minority. At the five-month senior course at NDC in 2014, for example, only seven of the 78 participating officers and civilian officials were from the U.S. The others came from 28 countries, including 12 from non-NATO countries around the world. NDC is a senior service college, and the U.S. Army and Marine Corps provide military education level 1 ("senior service college") credit to those attending. U.S. senior service colleges have included substantial numbers of international attendees in the past 13 years, many of which are funded through the State Department's IMET program, but the majority and the environment is a U.S.-dominated one.

As for operational assignments, few U.S. military personnel experience deployment in a coalition or multinational organization that is not structured around a U.S. military command and staffed with a majority of U.S. personnel. An exception is those U.S. personnel serving in United Nations peacekeeping missions such as those in the Sinai, Cyprus, or the Balkans. Such assignments could be encouraged as critical educational, training and broadening experiences. The most powerful incentive would be to include such experience as a criterion or credit toward promotion or command assignments.

Another way in which the United States could improve the multinational aspect of operations is to identify and deliberately plan to use specific expertise or skills that non-U.S. forces possess. The expertise of coalition partners such as Britain through its long conflict with Northern Ireland was usefully leveraged in reconciliation efforts in Afghanistan and Iraq, primarily through the person of the three-star general who led the reintegration cell in both MNF-I and ISAF. The national police or carabineri of Italy were useful in training Afghan police forces. Australia's Federal Police has a standing police advisory corps that has deployed to support stability and peacekeeping operations in over a half-dozen countries.[66]

The past 13 years revealed a chronic gap in U.S. police advisory and training capability. The United States has no national police force, and its efforts at building police capacity were less than optimal in both Iraq and Afghanistan. The effort was undertaken belatedly, with both civilian and military personnel that did not receive extensive training in the functional mission, the language, or the culture. There is a strong likelihood that the gap will persist in the current budget environment. The demand is likely to persist. A coalition effort to develop doctrine on policing models and training, and a plan to rely on coalition members' national police forces to conduct future training missions, might fill the chronic deficit of knowledge and capacity evident in Iraq and Afghanistan.

Finally, the internal transformations that the UK and Australian forces are pursuing can provide valuable lessons for the U.S. force despite the latter's much greater size and reach. The UK and Australian armies are both emphasizing task organization at lower echelons and cross-functional personnel. The Australian Army is currently reorganizing to build "a force first around a mission" rather than focusing battle groupings around battalions.[67] This reorganizational concept seeks to provide a more "flexible, modular combined arms teams" to better address contemporary threats.[68] Similarly, the British Army is pursuing an approach it calls "cross-organizational" aimed at maintaining a baseline of competence in combined-arms skills while ensuring a degree of adaptability.[69] The British are also emphasizing the importance of training the Reserve component of its force and better integrating them into the joint force by concentrating on "individuals, sub-units and formed units."[70]

---

[66] Australian Federal Police, *International Deployment Group Fact Sheet*, August 2008.

[67] Ryan, forthcoming, p. 6.

[68] The Australian Army, 2014, p. 35.

[69] UK Ministry of Defence, 2012, pp. 4–5.

[70] UK Ministry of Defence, 2012, pp. 4–5. See also pp. 3–5.

## Conclusion

This study suggests that recognizing the likely continuities between the recent past and the possible future will provide a hedge against unwise abandonment of hard-won innovations in practice and thought. While budget decisions necessarily force reductions in capacity, many of the capabilities developed over the past 13 years merit retention at smaller scale. Some of those capabilities warrant further investment of time or resources to ensure they are refined to perform better in the future. And some ongoing gaps, if not addressed, represent a risk to future mission success. Some of those gaps may be remedied by improving interagency, intergovernmental, and multinational coordination, which can yield important efficiencies. The United States may find that other partners are better suited not only to perform certain needed roles but to conceptualize and lead them.

The tendency at the present moment, as the large commitments in Iraq and Afghanistan have ended, may be to completely eliminate capabilities that were developed and dismiss as irrelevant lessons that were only partially formulated and disseminated. This rush to turn the page on the past 13 years may impose a heavy price, outweighing the presumed savings gained, when the next U.S. intervention occurs. As Feaver noted, "The United States has a cyclical tendency to follow too much expansion with too much retrenchment, and vice versa. Policymakers often overcompensate, at least in their rhetoric, for the actions of their predecessors. Successful policy must avoid this temptation, as it judiciously mixes opportunities for cost-saving cuts with continued commitments to extended security for the nation and its diverse interests."[71]

The continued official effort to develop lessons based on in-depth study of recent experience and the proposed joint concept for sustained land operations represent two important avenues for capturing requirements for future doctrine and other capability development. The ideas for theoretical development and institutional reform suggested by this

---

[71] Peter Feaver, ed., *Strategic Retrenchment and Renewal in the American Experience,* Strategic Studies Institute, Army War College, 2014, pp. 5–6.

analysis may aid further official concept development and experimentation and provide a basis for additional academic research. Finally, both the executive branch and the U.S. Congress may consider embracing a small set of educational and policy reforms that would improve civilian competence in national security strategy and increase critically needed civilian capability, the benefits of which would more than compensate for their relatively modest cost.

# References

"3-Star Details New Marine Crisis-Response Force," *Marine Corps Times*, April 21, 2014. As of July 16, 2014:
http://www.marinecorpstimes.com/article/20130421/
NEWS/304210007/3-star-details-new-Marine-crisis-response-force

Abizaid, John P., Rosa Brooks, et al., *Recommendation and Report of the Task Force on U.S. Drone Policy*, Stimson Center, June 2014.

American Academy of Diplomacy, *Forging a 21st Century Diplomatic Service for the United States Through Professional Education and Training*, Washington, D.C.: The Henry L. Stimson Center, 2011.

Amos, James F., *2014 Report to Congress on the Posture of the United States Marine Corps*, House Armed Services Committee, 2014.

Andres, Richard B., Craig Wills, and Thomas Griffith Jr., "Winning with Allies: The Strategic Value of the Afghan Model," *International Security*, Vol. 30, No. 3, Winter 2005/06, pp. 124–160.

Army Posture Statement, Testimony Before Senate Armed Services Committee, April 3, 2014.

The Australian Army, *Land Warfare Doctrine 1: The Fundamentals of Land Power 2014*, Australian Army Headquarters, 2014.

Australian Federal Police, *International Deployment Group Fact Sheet,* August 2008.

Barfield, Thomas, *Afghanistan: A Cultural and Political History*, Princeton, N.J.: Princeton University Press, 2012.

Bartholomees, J. Boone, "Theory of Victory," *Parameters*, Summer 2008.

Belasco, Amy, "Troop Levels in the Afghan and Iraq Wars, FY2001–FY2012: Cost and Other Potential Issues," Washington, D.C.: Congressional Budget Office, 2009.

Bensahel, Nora, Olga Oliker, Keith Crane, Rick Brennan, Jr, Heather S. Gregg, Thomas Sullivan, and Andrew Rathmell, *After Saddam: Prewar Planning and the Occupation of Iraq*, Santa Monica, Calif.: RAND Corporation, MG-642-A, 2008. As of September 23, 2014:
http://www.rand.org/pubs/monographs/MG642.html

Bensahel, Nora, Olga Oliker, and Heather Peterson, *Improving Capacity for Stabilization and Reconstruction Operations*, Santa Monica, Calif.: RAND Corporation, MG-852-OSD, 2009. As of September 23, 2014:
http://www.rand.org/pubs/monographs/MG852.html

Berman, Eric G., Matt Schroeder, and Jonah Leff, "Anti-Tank Guided Weapons," Research Note No. 16, *Small Arms Survey*, April 2012. As of August 7, 2014:
http://www.smallarmssurvey.org/fileadmin/docs/H-Research_Notes/SAS-Research-Note-16.pdf

Berman, Eric G., Matt Schroeder, and Jonah Leff, "Man-Portable Air Defence Systems (MANPADS)," Research Note No. 1, *Small Arms Survey*, 2011

Berzins, Janis, "Russia's New Generation Warfare in Ukraine: Implications for Latvian Defense Policy," *National Defence Academy of Latvia Policy Paper Number 02*, April 2014.

Betts, Richard K., "Is Strategy an Illusion?" *International Security*, Vol. 25, No. 2, 2000.

Biddle, Stephen D., "Afghanistan's Legacy: Emerging Lessons of an Ongoing War," *The Washington Quarterly*, Vol. 37, No. 2, 2014.

Biddle, Stephen D., "Allies, Airpower, and Modern Warfare: The Afghan Model in Afghanistan and Iraq," *International Security*, Vol. 30, No. 3, Winter 2005/06, pp. 161–176.

Biddle, Stephen, "Afghanistan and the Future of Warfare," *Foreign Affairs*, Vol. 82, No. 2, March–April 2003.

Biddle, Stephen D., *Afghanistan and the Future of Warfare: Implications for Army and Defense Policy*, Carlisle, Pa.: Strategic Studies Institute, U.S. Army War College, 2002.

Biddle, Stephen D., Jeffrey A. Friedman, and Jacob N. Shapiro, "Testing the Surge: Why Did Violence Decline in Iraq in 2007?" *International Security*, Vol. 37, No. 1, 2012.

Binnendijk, Hans, and Patrick M. Cronin, *Civilian Surge: Key to Complex Operations*, National Defense University, Washington, D.C.: Center for Technology and National Security Policy, 2009.

Bloomfield, Jr., Lincoln P., "Brave New World War," *Campaigning: Journal of the Department of Operational Art and Campaigning*, Joint Advanced Warfighting School, Summer 2006.

Boot, Max, *The Savage Wars of Peace: Small Wars and the Rise of American Power*, New York: Basic Books, 2002, pp. 281–335.

Brennan, Jr., Rick, Charles P. Ries, Larry Hanauer, Ben Connable, Terrence K. Kelly, Michael J. McNerney, Stephanie Young, Jason H. Campbell, and K. Scott McMahon, *Ending the U.S. War in Iraq*, Santa Monica, Calif.: RAND Corporation, RR-232-USFI, 2013. As of September 23, 2014: http://www.rand.org/pubs/research_reports/RR232.html

Brennan, Jr., Rick, Charles P. Ries, Larry Hanauer, Ben Connable, Terrence K. Kelly, Michael J. McNerney, Stephanie Young, Jason H. Campbell, and K. Scott McMahon, "Smooth Transitions? Lessons Learned from Transferring U.S. Military Responsibilities to Civilian Authorities in Iraq," Santa Monica, Calif.: RAND Corporation, RB-9749-USFI, 2013. As of September 23, 2014: http://www.rand.org/pubs/research_briefs/RB9749.html

Briscoe, Charles H., et al., *All Roads Lead to Baghdad: Army Special Operations Forces in Iraq*, New York: Paladin Press, 2007.

Briscoe, Charles H., et al., *Weapon of Choice: ARSOF in Afghanistan*, Fort Leavenworth, Kan.: Combat Studies Institute Press, 2003.

Bulos, Nabih, Patrick J. McDonnell, and Raja Abdulrahim, "ISIS Weapons Windfall May Alter Balance in Iraq, Syria Conflicts," *Los Angeles Times*, June 29, 2014. As of September 23, 2014: http://www.latimes.com/world/middleeast/ la-fg-iraq-isis-arms-20140629-story.html#page=1

Burk, James, *How 9/11 Changed Our Ways of War*, Stanford, Calif.: Stanford University Press, 2013.

Byman, Daniel, "Why Drones Work: The Case for Washington's Weapon of Choice," *Foreign Affairs*, Vol. 92, No. 4, 2013.

Byman, Daniel L., and Matthew C. Waxman, "Kosovo and the Great Air Power Debate," *International Security*, Vol. 24, No. 4, Spring 2000, pp. 5–38.

Caslen, Robert L., Jr., et al., "Security Cooperation Doctrine and Authorities: Closing the Gaps," *Joint Force Quarterly*, Vol. 71, No. 4.

Cassidy, Robert M., "Back to the Street Without Joy: Counterinsurgency Lessons from Vietnam and Other Small Wars," *Parameters*, Summer 2004.

Center for the Study of the Presidency & Congress, Project on National Security Reform, "Forging a New Shield," November 2008. As of August 14, 2014: http://www.thepresidency.org/programs/past-programs/ project-on-national-security-reform

Chandrasekaran, Rajiv, *Little America: The War Within the War for Afghanistan*, New York: Vintage Books, 2012.

Chivvis, Christopher S., and Jeffrey Martini, *Libya After Qaddafi: Lessons and Implications for the Future*, Santa Monica, Calif.: RAND Corporation, RR-577-SRF, 2014. As of September 23, 2014:
http://www.rand.org/pubs/research_reports/RR577.html

Chivvis, Christopher S., Olga Oliker, Andrew Liepman, Ben Connable, George Willcoxon, and William Young, "Initial Thoughts on the Impact of the Iraq War on U.S. National Security Structures," Santa Monica, Calif.: RAND Corporation, PE-111-OSD, 2014. As of September 23, 2014:
http://www.rand.org/pubs/perspectives/PE111.html

Chuter, Andrew, "5,000 Libyan MANPADS Secured: Some May Have Been Smuggled Out," Defense News, April 12, 2012. As of August 7, 2014:
http://www.defensenews.com/article/20120412/DEFREG04/304120002/
5-000-Libyan-MANPADS-Secured?odyssey=tab|topnews|text|FRONTPAGE

Clapper, James R., "Statement for the Record: Worldwide Threat Assessment of the U.S. Intelligence Community," Senate Select Committee on Intelligence, January 29, 2014.

Cleveland, Charles, "Command and Control of the Joint Commission Observer Program U.S. Army Special Forces in Bosnia," Strategy Research Project, United States Army War College, 2001. As of September 5, 2014:
http://oai.dtic.mil/oai/
oai?verb=getRecord&metadataPrefix=html&identifier=ADA391195

Cleveland, Charles T., and Stuart Farris, "Toward Strategic Landpower," *Army Magazine*, July 2013, pp. 20–23.

Cohen, Eliot A., "The Historical Mind and Military Strategy," *Orbis*, Fall 2005.

Cohen, Eliot A., *Supreme Command: Soldiers, Statesmen, and Leadership in Wartime*, New York: Free Press, 2002.

Cohen, Eliot A., "Constraints on America's Conduct of Small Wars," *International Security*, Vol. 9, No. 2, Fall 1984, pp. 178–179.

Cohen, Raphael S., "Just How Important Are 'Hearts and Minds' Anyway? Counterinsurgency Goes to the Polls," *Journal of Strategic Studies*, ahead of print, 2014, pp. 1–28.

Cohen, Raphael S., "A Tale of Two Manuals," *Prism*, Vol. 2, No. 1, 2010, pp. 87–100.

Colby, William, and James McCargar, *Lost Victory: A Firsthand Account of America's Sixteen-Year Involvement in Vietnam*, Chicago, Ill.: Contemporary Books, 1989.

Cole, Beth, "Guiding Principles for Stabilization and Reconstruction," *United States Army Peacekeeping and Stability Operations Institute*, United States Institute of Peace, 2009.

Collins, Joseph, "Planning Lessons from Afghanistan and Iraq." JFQ Forum 2nd Quarter, 2006, pp. 10–14.

Connable, Ben, Walter L. Perry, Christopher Paul, K. Scott McMahon, Erin York, and Todd Nichols, "Geospatially-Focused Socio-Cultural Analysis at the U.S. Central Command's Afghanistan-Pakistan Center: A Review of the Human Terrain Analysis Branch (HTAB)," unpublished RAND Corporation research, 2012.

Crane, Conrad, "Phase IV Operations: Where Wars Are Really Won," *Military Review*, May–June 2005.

Crane, Conrad, *Avoiding Vietnam, the U.S. Army's Response to Defeat in Southeast Asia*, U.S. Army War College, 2002.

Cronin, Audrey Kurth, "The 'War on Terrorism': What Does It Mean to Win?" *Journal of Strategic Studies*, Vol. 37, No. 2, 2014.

Cronin, Patrick M., *Restraint: Recalibrating American Strategy*, Washington, D.C.: Center for a New American Security, 2010.

Cumings, Bruce, *A Korean War: A History*, New York: Random House, 2010.

Davidson, Janine, "The Contemporary Presidency: Civil-Military Friction and Presidential Decision Making: Explaining the Broken Dialogue," *Presidential Studies Quarterly*, Vol. 43, No. 1, 2013.

Davis, Paul K., *Military Transformation? Which Transformation, and What Lies Ahead?* Santa Monica, Calif.: RAND Corporation, RP-1413, 2010. As of September 25, 2014:
http://www.rand.org/pubs/reprints/RP1413.html

Davis, Lynn E., Michael J. McNerney, James S. Chow, Thomas Hamilton, Sarah Harting, and Daniel Byman, *Armed and Dangerous? UAVs and U.S. Security*, Santa Monica, Calif.: RAND Corporation, RR-449-RC, 2014. As of September 25, 2014:
http://www.rand.org/pubs/research_reports/RR449.html

Davis, Lynn E., and Jeremy Shapiro, eds., *The U.S. Army and the New National Security Strategy*, Santa Monica, Calif.: RAND Corporation, MR-1657-A, 2003. As of September 25, 2014:
http://www.rand.org/pubs/monograph_reports/MR1657.html

Davis, Paul K., and Peter A. Wilson, *Looming Discontinuities in U.S. Military Strategy and Defense Planning: Colliding RMAs Necessitate a New Strategy*, Santa Monica, Calif.: RAND Corporation, OP-326-OSD, 2011. As of September 25, 2014:
http://www.rand.org/pubs/occasional_papers/OP326.html

Destler, I. M., "National Security Advice to US Presidents: Some Lessons from Thirty Years," *World Politics*, Vol. 29, No. 2, 1977, pp. 143–176.

Dobbins, James, "Retaining the Lessons of Nationbuilding," in *Commanding Heights: Strategic Lessons from Complex Operations*, Washington, D.C.: NDU Press, 2010.

Dobbins, James, Seth G. Jones, Keith Crane, Andrew Rathmell, Brett Steele, Richard Teltschik, and Anga R. Timilsina, *The UN's Role in Nation-Building: From the Congo to Iraq*, Santa Monica, Calif.: RAND Corporation, MG-304-RC, 2005. As of September 25, 2014:
http://www.rand.org/pubs/monographs/MG304.html

Dobbins, James, Seth G. Jones, Benjamin Runkle, and Siddharth Mohandas, *Occupying Iraq: A History of the Coalition Provisional Authority*, Santa Monica, Calif.: RAND Corporation, MG-847-CC, 2009. As of September 25, 2014:
http://www.rand.org/pubs/monographs/MG847.html

Dobbins, James, John G. McGinn, Keith Crane, Seth G. Jones, Rollie Lal, Andrew Rathmell, Rachel Swanger, and Anga Timilsina, *America's Role in Nation-Building: From Germany to Iraq*, Santa Monica, Calif.: RAND Corporation, MR-1753-RC, 2003. As of September 25, 2014:
http://www.rand.org/pubs/monograph_reports/MR1753.html

Dunstan, Simon, *Vietnam Choppers, Helicopters in Battle 1950–1975*, London: Osprey Publishing, 2003.

Edwards, George C., III, and Stephen J. Wayne, *Presidential Leadership: Politics and Policy Making*, New York: Worth Publishers, 1999.

"Emergence of Smart Bombs," *Air Force Magazine*, 2014.

Feaver, Peter, ed., *Strategic Retrenchment and Renewal in the American Experience*, Strategic Studies Institute, Army War College, 2014.

Feaver, Peter, "The Right to Be Right: Civil-Military Relations and the Iraq Surge Decision," *International Security*, Vol. 35, No. 4, Spring 2011.

Feaver, Peter, "Crisis as Shirking: An Agency Theory Explanation of the Souring of American Civil-Military Relations," *Armed Forces & Society*, Vol. 24, No. 3, 1998.

Feaver, Peter, and Stephen Biddle, "Assessing Strategic Choices in the War on Terror," in James Burk, *How 9/11 Changed Our Ways of War*, Stanford, Calif.: Stanford University Press, 2013.

Feaver, Peter, and William Inboden, "A Strategic Planning Cell on National Security," in Drezner, Daniel W., ed., *Avoiding Trivia: The Role of Strategic Planning in American Foreign Policy*, Brookings Institution Press, 2009.

Feaver, Peter, and Richard H. Kohn, eds., *Soldiers and Civilians: The Civil-Military Gap and American National Security*, Cambridge, Mass.: MIT Press, 2001.

Fitzgerald, Ben, and Kelley Sayler, *Creative Disruption: Technology, Strategy, and the Future of the Global Defense Industry*, Washington, D.C.: Center for a New American Security, 2014.

Flournoy, Michele, and Janine Davidson, "Obama's New Global Posture: The Logic of U.S. Foreign Deployments," *Foreign Affairs,* Vol. 91, 2012, p. 64.

Flynn, Michael, "Annual Threat Assessment," Statement before the Senate Armed Services Committee, February 11, 2014.

Flynn, Michael T., et al., "Fixing Intel: Making Intelligence Relevant in Afghanistan," Washington, D.C.: Center for a New American Security, January 4, 2010.

Gates, Robert M., *Duty: Memoirs of a Secretary at War,* New York: Alfred A. Knopf, 2014.

Gates, Robert M., "Speech to the United States Military Academy (West Point, NY)," February 25, 2011. As of August 29, 2014:
http://www.defense.gov/speeches/speech.aspx?speechid=1539

*General Framework Agreement for Peace in Bosnia and Herzegovina,* a.k.a. "Dayton Accord," 1995, Annex IA.

Gentile, Gian, *Wrong Turn: America's Deadly Embrace of Counterinsurgency,* New York: The New Press, 2013.

Gentile, Gian P., "Time for the Deconstruction of Field Manual 3–24," *Joint Force Quarterly,* Vol. 58, No. 3, 2010, pp. 116–117.

*Global Trends 2030: Alternative Worlds: A Publication of the National Intelligence Council,* Washington, D.C.: National Intelligence Council, December 2012.

Gordon, John, and Pirnie, Bruce, "Everybody Wanted Tanks, Heavy Forces in Operation Iraqi Freedom," *Joint Force Quarterly,* Issue 39, 2005.

Gray, Colin S., *The Strategy Bridge,* New York: Oxford University Press, 2010.

Gray, Colin S., "Irregular Enemies and the Essence of Strategy: Can the American Way of War Adapt?" Carlisle, Pa.: Strategic Studies Institute, 2006.

Gray, Colin S., *Defining and Achieving Decisive Victory,* Carlisle, Pa.: U.S. Army War College, Strategic Studies Institute, 2002.

Greentree, Todd R., "Lessons from Limited Wars: A War Examined, Afghanistan," *Parameters,* Vol. 43, No. 3, 2013, p. 93.

Hastings, Max, *The Korean War,* New York: Simon and Schuster, 1987.

Hastings, Max, *Overlord, D-Day and the Battle for Normandy,* New York: Simon and Schuster, 1984.

HistoryShots, "U.S. Army Divisions in World War II," 2014. As of August 6, 2014:
http://www.historyshots.com/usarmy/backstory.cfm

Hoffman, Francis G., "Enhancing America's Strategic Competency," in Alan Cromartie, ed., *Liberal Wars,* London: Routledge, forthcoming.

Hoffman, Francis G., *Conflict in the 21st Century: The Rise of Hybrid Wars*, Arlington, Va.: Potomac Institute for Policy Studies, 2007.

Hosmer, Stephen T., *Psychological Effects of U.S. Air Operations in Four Wars, 1941–1991: Lessons for U.S. Commanders*, Santa Monica, Calif.: RAND Corporation, MR-576-AF, 1996. As of September 25, 2014:
http://www.rand.org/pubs/monograph_reports/MR576.html

House, Jonathan, *Combined Arms Warfare in the Twentieth Century*, Lawrence, Kan.: University of Kansas, 2001.

International Institute for Strategic Studies, *The Military Balance 2014*, 2014. As of August 21, 2014:
http://www.iiss.org/en/publications/military-s-balance

ISAF, "After Action Review Report: NATO-Afghanistan Transformation Task Force (NATTF)," HQ ISAF, Kabul, 2013.

Isely, Jeter A., and Philip A. Crowl, *The Marines and Amphibious War*, Princeton, N.J.: Princeton University Press, 1951.

Johnson, David E., *Hard Fighting: Israel in Lebanon and Gaza*, Santa Monica, Calif.: RAND Corporation, MG-1085-A/AF, 2011. As of September 25, 2014:
http://www.rand.org/pubs/monographs/MG1085.html

Johnson, David E., *Military Capabilities for Hybrid War: Insights from the Israel Defense Forces in Lebanon and Gaza*, Santa Monica, Calif.: RAND Corporation, OP-285-A, 2010. As of September 25, 2014:
http://www.rand.org/pubs/occasional_papers/OP285.html

Johnson, David E., *Learning Large Lessons: The Evolving Roles of Ground and Air Power in the Post-Cold War Era*, Santa Monica, Calif.: RAND Corporation, MG-405-1-AF, 2007. As of September 25, 2014:
http://www.rand.org/pubs/monographs/MG405-1.html

Johnson, David E., M. W. Markel, and Brian Shannon, *The 2008 Battle Of Sadr City: Reimagining Urban Combat*, Santa Monica, Calif.: RAND Corporation, RR-160-A, 2013. As of September 25, 2014:
http://www.rand.org/pubs/research_reports/RR160.html

Johnson, Matthew, "The Growing Relevance of Special Operations Forces in U.S. Military Strategy," *Comparative Strategy*, Vol. 25, No. 4, 2006.

Joint and Coalition Operational Analysis, *Decade of War, Volume 1: Enduring Lessons from the Past Decade of Operations*, June 15, 2012. As of August 6, 2014:
http://blogs.defensenews.com/saxotech-access/pdfs/decade-of-war-lessons-learned.pdf

Joint IED Defeat Organization, "About JIEDDO," undated. As of September 2, 2014:
https://www.jieddo.mil/about.aspx

Kaplan, Fred, *The Insurgents: David Petraeus and the Plot to Change the American Way of War*, New York, N.Y.: Simon & Schuster, 2013.

Kelly, Terrence K., James Dobbins, David A. Shlapak, David C. Gompert, Eric Heginbotham, Peter Chalk, and Lloyd Thrall, *The U.S. Army in Asia, 2030–2040*, Santa Monica, Calif.: RAND Corporation, RR-474-A, 2014. As of September 25, 2014:
http://www.rand.org/pubs/research_reports/RR474.html

Kelly, Terrence K., Nora Bensahel, and Olga Oliker, *Security Force Assistance in Afghanistan: Identifying Lessons for Future Efforts*, Santa Monica, Calif.: RAND Corporation, MG-1066-A, 2011. As of September 25, 2014:
http://www.rand.org/pubs/monographs/MG1066.html

Kelly, Terence K., Ellen E. Tunstall, Thomas S. Szayna, and Deanna Weber Prine, *Stabilization and Reconstruction Staffing: Developing U.S. Civilian Personnel Capabilities*, Santa Monica, Calif.: RAND Corporation, MG-580-RC, 2008. As of September 25, 2014:
http://www.rand.org/pubs/monographs/MG580.html

Kennan, George, *Policy Planning Staff Memorandum 269*, Records of the National Security Council RG 273, NSC 10/2, Washington, D.C., May 4, 1948.

Kershner, Isabel, "Missile from Gaza Hits School Bus," *New York Times*, April 7, 2011. As of August 7, 2014:
http://www.nytimes.com/2011/04/08/world/middleeast/08gaza.html?_r=0

Kervkliet, Benedict J., *The Huk Rebellion: A Study of Peasant Revolt in the Philippines*, University of California Press, 1977.

Krepinevich, Andrew F., *The Army and Vietnam*, Baltimore, Md.: Johns Hopkins University Press, 1986.

Krepinevich, Andrew F., Jr., and Barry D. Watts, *Regaining Strategic Competence*, Washington, D.C.: Center for Strategic and Budgetary Assessments, 2009.

Lamb, Christopher J., James Douglas Orton, Michael C. Davies, and Theodore T. Pikulsky, "The Way Ahead for Human Terrain Teams," *Joint Force Quarterly*, Vol. 70, No. 3, 2013, pp. 21–29.

Lieven, Anatol, and John Hulsman, *Ethical Realism*, New York: Random House, 2009.

Lister, Charles, "American Anti-Tank Weapons Appear in Syrian Rebel Hands," *Huffington Post*, April 9, 2014. As of August 7, 2014:
http://www.huffingtonpost.com/charles-lister/
american-anti-tank-weapon_b_5119255.html

Madden, Dan, Bruce R. Nardulli, Dick Hoffmann, Michael Johnson, Fred T. Krawchuk, John E. Peters, Linda Robinson, and Abby Doll, *Toward Operational Art in Special Warfare*, Santa Monica, Calif.: RAND Corporation, forthcoming.

Mahnken, Thomas, *Technology and the American Way of War Since 1945*, New York: Columbia University Press, 2010.

Manwaring, Max G., *Gangs, Pseudo-Militaries, and Other Modern Mercenaries: New Dynamics in Uncomfortable Wars*, Norman, Okla.: University of Oklahoma, 2010.

Mandelbaum, Michael, *The Frugal Superpower: America's Global Leadership in a Cash-Strapped Era*, New York: PublicAffairs, 2011.

Markel, M. Wade, Henry A. Leonard, Charlotte Lynch, Christina Panis, Peter Schirmer, and Carra S. Sims, *Developing U.S. Army Officers' Capabilities for Joint, Interagency, Intergovernmental, and Multinational Environments*, Santa Monica, Calif.: RAND Corporation, MG-990-A, 2011. As of September 25, 2014: http://www.rand.org/pubs/monographs/MG990.html

Marquis, Jefferson P., Jennifer D. P. Moroney, Justin Beck, Derek Eaton, Scott Hiromoto, David R. Howell, Janet Lewis, Charlotte Lynch, Michael J. Neumann, and Cathryn Quantic Thurston, *Developing an Army Strategy for Building Partner Capacity for Stability Operations*, Santa Monica, Calif.: RAND Corporation, MG-942-A, 2010. As of September 25, 2014: http://www.rand.org/pubs/monographs/MG942.html

Mazarr, Michael J., *The Revolution in Military Affairs: A Framework for Defense Planning*, Army War College, Strategic Studies Institute, Carlisle Barracks, Pa., 1994.

McChrystal, Stanley A., *My Share of the Task: A Memoir*, New York: Portfolio/Penguin, 2013.

McGurk, Brett, State Department Deputy Assistant Secretary for Iraq and Iran, testimony for Senate Foreign Relations Committee, July 23, 2014.

McMaster, H. R., "Decentralization vs Centralization," in Thomas Donnelly and Frederick W. Kagan, *Lessons for a Long War: How America Can Win on New Battlefields*, Washington, D.C.: AEI, 2010.

McMaster, H. R., "Effective Civilian-Military Planning," in Michael Miklaucic, ed., *Commanding Heights: Strategic Lessons from Complex Operations*, NDU Press, 2010.

McMaster, H. R., "On War: Lessons to Be Learned," *Survival*, Vol. 50, No. 1, 2008.

McRaven, William H., "Testimony to the U.S. Senate on the Department of Defense Authorization of Appropriations for Fiscal Years 2015 and the Future Years Defense Program," U.S. Senate Armed Services Committee, March 11, 2014.

Metz, Steve, *Strategic Landpower Task Force Research Report*, Strategic Studies Institute, October 3, 2013.

Miller, Greg, "ISIS Rapidly Accumulating Cash, Weapons, U.S. Intelligence Officials Say," *Washington Post*, June 24, 2014. As of September 25, 2014: http://www.washingtonpost.com/world/national-security/ isis-rapidly-accumulating-cash-weapons-us-intelligence-officials-say/2014/06/24/ bd050770-fbda-11e3-932c-0a55b81f48ce_story.html

Miller, Paul D., "Armed State Building: Confronting State Failure, 1898–2012," Cornell University Press, 2013.

Miller, Paul D., "The Contemporary Presidency: Organizing the National Security Council: I Like Ike's," *Presidential Studies Quarterly*, Vol. 43, No. 3, 2013, pp. 592–606.

Morgenthau, Hans J., and Kenneth W. Thompson, *Politics Among Nations: The Struggle for Power and Peace*, New York: Knopf, 1978.

Murdock, Clark A., et al., *Beyond Goldwater-Nichols: Defense Reform for a New Strategic Era,* Phase II Report, Washington, D.C.: Center for Strategic and International Studies, 2006.

Murdock, Clark A., et al., *Beyond Goldwater-Nichols: Defense Reform for a New Strategic Era,* Phase 1 Report, Washington, D.C.: Center for Strategic and International Studies, March 2004.

Nagl, John A., "Constructing the Legacy of Field Manual 3-24," *Joint Forces Quarterly*, Vol. 58, No. 3, 2010, pp. 118–120.

Nagl, John, *Learning to Eat Soup with a Knife: Counterinsurgency Lessons from Malaya and Vietnam*, Chicago, Ill.: University of Chicago, 2002.

Nardulli, Bruce R., Walter L. Perry, Bruce R. Pirnie, John Gordon IV, and John G. McGinn, *Disjointed War: Military Operations in Kosovo, 1999,* Santa Monica, Calif.: RAND Corporation, MR-1406-A, 2002. As of September 25, 2014: http://www.rand.org/pubs/monograph_reports/MR1406.html

National Archives, "Organization of the Office of Strategic Services (Record Group 226)," undated. As of September 4, 2014: http://www.archives.gov/research/military/ww2/oss/personnel-files.html

National Intelligence Council, *Global Trends 2030: Alternative Worlds*, NIC 2012-001, December 2012.

National Security Presidential Directive 44 (NSPD-44), *Management of Interagency Efforts Concerning Reconstruction and Stabilization*, Washington, D.C., December 7, 2005. As of September 25, 2014: http://fas.org/irp/offdocs/nspd/nspd-44.html

National WWII Museum, "By the Numbers: The U.S. Military—U.S. Military Personnel (1939–1945)," undated. As of August 6, 2014: http://www.nationalww2museum.org/learn/education/for-students/ww2-history/ ww2-by-the-numbers/us-military.html

NATO, "Senior Civilian Representative Report: A Comprehensive Approach Lessons Learned in Afghanistan," July 15, 2010.

Navy Department Library, "U.S. Navy Personnel in World War II: Service and Casualty Statistics," *Annual Report, Navy and Marine Corps Military Personnel Statistics*, June 30, 1964. As of August 6, 2014:
http://www.history.navy.mil/library/online/ww2_statistics.htm

Neustadt, Richard E., *Presidential Power*, New York: New American Library, 1960.

O'Hanlon, Michael, "Flawed Masterpiece," *Foreign Affairs*, Vol. 81, No. 3, May–June 2002.

O'Hanlon, Michael, *Technological Change and the Future of Warfare*, Washington, D.C.: Brookings Institution Press, 2000.

Obama, Barack, speech, West Point, N.Y., May 29, 2014. As of September 25, 2014:
http://www.nytimes.com/2014/05/29/us/politics/
transcript-of-president-obamas-commencement-address-at-west-point.html?_r=0

Ogorkiewicz, Richard M., *Armored Forces, A History of Armored Forces and Their Vehicles*, New York: Arco Publishing, 1970.

Olsen, Howard, and John Davis, "Training U.S. Army Officers for Peace Operations: Lessons from Bosnia," United States Institute of Peace Special Report, October 29, 1999.

"Panetta: 'My Mission Has Always Been to Keep the Country Safe,'" National Public Radio, February 3, 2013. As of August 14, 2014:
http://www.npr.org/2013/02/03/170970194/
panetta-my-mission-has-always-been-to-keep-the-country-safe?ft=1&f=

Paul, Christopher, Colin P. Clarke, Beth Grill, Stephanie Young, Jennifer D. P. Moroney, Joe Hogler, and Christine Leah, *What Works Best When Building Partner Capacity and Under What Circumstances?* Santa Monica, Calif.: RAND Corporation, MG-1253/1-OSD, 2013. As of September 25, 2014:
http://www.rand.org/pubs/monographs/MG1253z1.html

Phillips, R. Cody, *Operation Just Cause: The Incursion into Panama*, Washington, D.C.: Center of Military History, 2006.

Posen, Barry R., "The Case for Restraint," *The American Interest*, Vol. 3, No. 1, 2007.

Posen, Barry R., *The Sources of Military Doctrine*, Ithaca, N.Y.: Cornell University Press, 1984.

Presidential Decision Directive 56 (PDD-56), *Managing Complex Contingency Operations*, White Paper, May 1997. As of September 25, 2014:
http://fas.org/irp/offdocs/pdd56.htm

Project on National Security Reform, *Forging a New Shield*, 2008.

Quadrennial Defense Review, 2014. As of September 12, 2014:
http://www.defense.gov/pubs/2014_Quadrennial_Defense_Review.pdf

Ricks, Thomas E., *Fiasco: The American Military Adventure in Iraq*, New York: Penguin Press, 2006.

Ripley, Tim, *The Wehrmacht: The German Army in World War II, 1939–1945*, Routledge, 2014, pp 16–18.

Robinson, Linda, *One Hundred Victories: Special Ops and the Future of American Warfare*, New York: PublicAffairs, 2013. As of September 25, 2014:
http://www.rand.org/pubs/commercial_books/CB535.html

Robinson, Linda, *Tell Me How This Ends: General David Petraeus and the Search for a Way Out of Iraq*, New York: PublicAffairs, 2008.

Robinson, Linda, "The End of El Salvador's War," *Survival*, Vol. 33, September/October 1991.

Rosen, Stephen P., "New Ways of War: Understanding Military Innovation," *International Security*, Vol. 13, No. 1, Summer 1988, pp. 134–168.

Rosenau, William, *Special Operations Forces and Elusive Enemy Ground Targets: Lessons from Vietnam and the Persian Gulf War*, Santa Monica, Calif.: RAND Corporation, MR-1408-AF, 2001. As of September 23, 2014:
http://www.rand.org/pubs/monograph_reports/MR1408.html

Ryan, Mick, "After Afghanistan: A Small Army and the Strategic Employment of Land Power," *Security Challenges,* forthcoming, Fall 2014.

Schadlow, Nadia, "Competitive Engagement: Upgrading America's Influence," *Orbis*, Vol. 57, No. 4, 2013, pp. 501–515.

Schadlow, Nadia, "War and the Art of Governance," *Parameters*, Vol. 33, No. 3, 2003.

Schroeder, Matt, "Appendix 14A: Global Efforts to Control MANPADS," in Stockholm International Peace Research Institute, *SIPRI Yearbook 2007: Armaments, Disarmament and International Security*, Oxford University Press.

Schuber, Frank, et al., *The Whirlwind War*, Washington: Center of Military History, 1995.

Scoville, Thomas W., *Reorganizing for Pacification Support*, Washington, D.C.: Center of Military History, U.S. Army, 1982, Chapter 5. As of September 25, 2014:
http://www.history.army.mil/books/Pacification_Spt/Ch5.htm

Seals, Bob, "The 'Green Beret Affair': A Brief Introduction," Military History Online, November 24, 2007. As of August 6, 2014:
http://www.militaryhistoryonline.com/20thcentury/articles/greenberets.aspx

Seddon, Max, "Documents Show How Russia's Troll Army Hit America," *BuzzFeed World*, June 2, 2014. As of August 8, 2014:
http://www.buzzfeed.com/maxseddon/
documents-show-how-russias-troll-army-hit-america

Serafino, Nina, "Peacekeeping/Stabilization and Conflict Transitions: Background and Congressional Action on the Civilian Response/Reserve Corps and Other Civilian Stabilization and Reconstruction Capabilities," Congressional Research Service, October 2, 2012. As of August 8, 2014:
http://fas.org/sgp/crs/natsec/RL32862.pdf

Simmons, Anna, *21st Century Cultures of War: Advantage Them,* Carlisle, Pa.: Foreign Policy Research Institute, April 2013.

Smith, Rupert, *The Utility of Force: The Art of War in the Modern World*, New York: Random House, 2008.

SOCOM Lessons Learned Operational and Strategic Studies Branch, "Special Operations Joint Task Force-Afghanistan (SOJTF-A): From Concept to Execution . . . The First Year," 2013.

Spirtas, Michael, Jennifer D. P. Moroney, Harry J. Thie, Joe Hogler, and Durell Young, *Department of Defense Training for Operations with Interagency, Multinational, and Coalition Partners*, Santa Monica, Calif.: RAND Corporation, MG-707-OSD, 2008. As of September 25, 2014:
http://www.rand.org/pubs/monographs/MG707.html

Stockholm International Peace Research Institute (SIPRI), SIPRI Arms Transfers Database, 2014. As of August 21, 2014:
http://www.sipri.org/databases/armstransfers

Strachan, Hew, *The Direction of War: Contemporary Strategy in Historical Perspective*, New York: Cambridge University Press, 2013.

Summers, Harry G., *On Strategy: A Critical Analysis of the Vietnam War*, New York: Random House, 1995.

Szayna, Thomas S., Derek Eaton, James E. Barnett, Brooke Stearns Lawson, Terrence K. Kelly, and Zachary Haldeman, *Integrating Civilian Agencies in Stability Operations*, Santa Monica, Calif.: RAND Corporation, MG-801-A, 2009. As of September 25, 2014:
http://www.rand.org/pubs/monographs/MG801.html

Szayna, Thomas S., Derek Eaton, and Amy Richardson, *Preparing the Army for Stability Operations: Doctrinal and Interagency Issues*, Santa Monica, Calif.: RAND Corporation, MG-646-A, 2007. As of September 25, 2014:
http://www.rand.org/pubs/monographs/MG646.html

Szayna, Thomas S., Derek Eaton, Stephen Watts, Joshua Klimas, James T. Quinlivan, and James C. Crowley, *Assessing Alternatives for Full Spectrum Operations and Security Force Assistance: Specialized vs. Multipurpose Army Forces*, unpublished RAND Corporation research, 2013.

Szayna, Thomas S., Kevin F. McCarthy, Jerry M. Sollinger, Linda J. Demaine, Jefferson P. Marquis, and Brett Steele, *The Civil-Military Gap in the United States: Does It Exist, Why, and Does It Matter?* Santa Monica, Calif.: RAND Corporation, MG-379-A, 2007. As of September 25, 2014:
http://www.rand.org/pubs/monographs/MG379.html

Szayna, Thomas S., Angela O'Mahony, Jennifer Kavanagh, Stephen Watts, Bryan A. Frederick, Tova C. Norlen, and Phoenix Voorhies, *Conflict Trends and Conflict Drivers: An Empirical Assessment of Historical Conflict Patterns and Future Conflict Projections*, unpublished RAND Corporation research, 2013.

U.S. Army, U.S. Marine Corps, and U.S. Special Operations Command, "Strategic Landpower: Winning the Clash of Wills," white paper, May 2013. As of September 12, 2014:
http://www.tradoc.army.mil/FrontPageContent/Docs/
Strategic%20Landpower%20White%20Paper.pdf

U.S. Army Special Operations Command, *Casebook on Insurgency and Revolutionary Warfare: 23 Summary Accounts*, Assessing Revolutionary and Insurgent Strategies, undated. As of September 25, 2014:
http://www.soc.mil/ARIS/ARIS.html

U.S. Army Special Operations Command, "USASOC Silent Quest Facilitated War Game 14-1 Executive Summary and Final Report," 2014.

U.S. Army Special Operations Command, "ARSOF 2022," 2013. As of September 25, 2014:
http://www.soc.mil/Assorted%20Pages/ARSOF2022_vFINAL.pdf

U.S. Army Special Operations Command, *Human Factors Considerations of Undergrounds in Insurgencies, 2d Edition, 2013*, Assessing Revolutionary and Insurgent Strategies, January 25, 2013. As of September 25, 2014:
http://www.soc.mil/ARIS/ARIS.html

U.S. Army Special Operations Command, *Undergrounds in Insurgent, Revolutionary, and Resistance Warfare, 2d Edition*, Assessing Revolutionary and Insurgent Strategies, January 25, 2013. As of September 25, 2014:
http://www.soc.mil/ARIS/ARIS.html

U.S. Army Special Operations Command, *Casebook on Insurgency and Revolutionary Warfare, Volume II 1962 – 2009*, Assessing Revolutionary and Insurgent Strategies, April 27, 2012. As of September 25, 2014:
http://www.soc.mil/ARIS/ARIS.html

U.S. Department of the Army, *TRADOC Pamphlet 525-3-1, The U.S. Army Operating Concept: Win in a Complex World*, Washington, D.C.: Government Printing Office, October 7, 2014.

U.S. Department of the Army, *Field Manual 3-24: Insurgencies and Countering Insurgencies*, Washington, D.C.: Government Printing Office, May 2014.

U.S. Department of the Army, *TRADOC Pam 525-8-5, U.S. Army Functional Concept for Engagement, 2014,* Washington, D.C.: Government Printing Office, February 2014.

U.S. Department of the Army, *Army Posture Statement 2014,* 2014.

U.S. Department of the Army, *Army Strategic Planning Guidance 2013,* 2013.

U.S. Department of the Army, *U.S. Army Doctrine Reference Publication 3-05, Special Operations,* Headquarters, Washington, D.C.: Government Printing Office, August 31, 2012.

U.S. Department of the Army, *The U.S. Army-Marine Corps Counterinsurgency Field Manual, U.S. Army Field Manual No. 3-24, U.S. Marine Corps Warfighting Publication No. 3-33.5,* Chicago, Ill.: University of Chicago Press, 2007.

U.S. Department of the Army, *The Brigade Combat Team FM 3-90.6,* Washington, D.C.: Government Printing Office, August 2006.

U.S. Department of Defense, *Joint Publication 1-02, Department of Defense Dictionary of Military and Associated Terms: 8 November 2010 (As Amended Through 16 July 2014),* Washington, D.C.: Government Printing Office, July 2014.

U.S. Department of Defense, Joint Publication 1, *Doctrine of the Armed Forces of the United States,* JP 1-6, 2013.

U.S. Department of Defense, *Capstone Concept for Joint Operations: Joint Force 2020,* Washington, D.C.: Government Printing Office, 2012.

U.S. Department of Defense, Joint Publication 3-07, *Stability Operations,* 2011.

U.S. Department of Defense, *Report of the Defense Science Board Task Force on: Understanding Human Dynamics,* Washington, D.C.: Office of the Under Secretary of Defense for Acquisition, Technology, and Logistics, 2009.

U.S. Department of the Navy, *Expeditionary Force 21,* Washington, D.C.: Department of the Navy, March 4, 2014, p. 24.

U.S. Department of the Navy, *Marine Corps Operations, MCDP 1-0,* August 9, 2011.

U.S. Department of State, "National Security Decision Directive (NSDD) 38," undated. As of August 6, 2014:
http://www.state.gov/m/pri/nsdd/

U.S. Department of State, "Occupation and Reconstruction of Japan, 1945–52," Office of the Historian, undated. As of August 6, 2014:
https://history.state.gov/milestones/1945-1952/japan-reconstruction

U.S. Department of State, "MANPADS: Combating the Threat to Global Aviation from Man-Portable Air Defense Systems," July 27, 2011. As of August 7, 2014:
http://www.state.gov/t/pm/rls/fs/169139.htm

U.S. Department of State, *U.S. Government Counterinsurgency Guide 2009*, Department of State, Washington, D.C.: Government Printing Office, 2009.

U.S. Government Accountability Office, *Further Improvements Needed in U.S. Efforts to Counter Threats from Man-Portable Air Defense Systems*, May 13, 2004.

U.S. Joint Forces Command, *Joint Operations Environment (JOE) 2010*, Joint Futures Group, Norfolk, Va., February 18, 2010.

Ucko, David, "Critics Gone Wild: Counterinsurgency as the Root of All Evil," *Small Wars & Insurgencies*, Vol. 25, No. 1, 2014.

UK Ministry of Defense, *Future Land Operating Concept: Joint Concept Note 2/12*, The Development, Concepts and Doctrine Centre, 2012.

Under Secretary of Defense (Comptroller), "FY2015 Counterterrorism Partnerships Fund and the European Reassurance Initiative," undated. As of September 25, 2014:
http://comptroller.defense.gov/budgetmaterials/budget2015.aspx

Under Secretary of Defense (Comptroller), "FY2015 DoD Overseas Contingency Operations (OCO) Budget Amendment," undated. As of September 25, 2014:
http://comptroller.defense.gov/budgetmaterials/budget2015.aspx

United Kingdom Ministry of Defence, *Global Strategic Trends—Out to 2045, Fifth Edition*, Strategic Trends Programme, 2014.

United Nations Security Council, *Report of the Monitoring Group on Somalia Submitted in Accordance with Resolution 1853 (2008)*.

USAID-USSOCOM, *Joint Sahel Project: Development Game After Action Report AAR*, May 27, 2014.

USASOC—*see* U.S. Army Special Operations Command.

*Vietnam Studies: U.S. Army Special Forces, 1961–1971*, Center for Military History Publication 90-23, Washington, D.C.: Department of the Army, 1989.

Vincent, Douglas, G., *Being Human Beings: The Domains and a Human Realm*, Strategy Research Project, U.S. Army War College, March 2013.

Von Bismarck, Otto, *Bismarck: The Man and the Statesman*, Volume 1, New York: Cosimo, Inc., 2005.

Walt, Stephen M., *Taming American Power: The Global Response to U.S. Primacy*, New York: W. W. Norton, 2005.

Watts, Stephen, Jason H. Campbell, Patrick B. Johnston, Sameer Lalwani, and Sarah H. Bana, *Countering Others' Insurgencies: Understanding U.S. Small-Footprint Interventions in Local Context*, Santa Monica, Calif.: RAND Corporation, RR-513-SRF, 2014. As of September 25, 2014:
http://www.rand.org/pubs/research_reports/RR513.html

Weigley, Russell F., *The American Way of War: A History of United States Military Strategy and Policy*, Bloomington, Ind.: Indiana University Press, 1977.

Wentz, Larry, *Lessons from Bosnia: The IFOR Experience*, Washington, D.C.: DoD CCRP NDU, 1997.

Wezeman, Siemon, et al., "International Arms Transfers," *SIPRI Yearbook 2007: Armaments, Disarmament and International Security*, Oxford: Oxford University Press, 2007.

Williamson, Murray, and Allan R. Millett, *A War to Be Won: Fighting the Second World War*, Cambridge, Mass.: Harvard University Press, 2009.

Williamson, Murray, and Allan R. Millett, *Military Innovation in the Interwar Period*, Cambridge University Press, 1998.

Wilson, Clay, "Network Centric Warfare, Background and Oversight Issues for Congress," Washington, D.C.: Congressional Research Service, June 2, 2004. As of August 6, 2014:
http://www.fas.org/man/crs/RL32411.pdf

Woodward, Bob, *Obama's Wars,* New York: Simon & Schuster, 2011.

Woodward, Bob, *Plan of Attack*, New York: Simon & Schuster, 2004.

Zegart, Amy, *Flawed by Design: The Evolution of the CIA, JCS, and NSC*, Stanford, Calif.: Stanford University Press, 1999.